TABLE OF CONTENTS

Foreword 3

Organizing Committee 7

Executive Summary and Key Assessments 9

Introduction and Overview of the Workshop 17

The Vision for Space Exploration and the Role of Science 21

The Global Exploration Strategy 23

The Lunar Exploration Architecture 27

Assessment of Potential Science Activities 31

Findings of the Workshop 35

    Astrophysics 36

    Earth Science 39

    Heliophysics 41

    Planetary Protection 43

    Planetary Science 46

Outreach Messages and Highlights of the Workshop 53

Concluding Statement 57

Appendices: Science Subcommittee Workshop Findings 59

    Appendix 1: Astrophysics 59

    Appendix 2: Earth Science 67

    Appendix 3: Heliophysics 83

    Appendix 4: Planetary Protection 91

    Appendix 5: Planetary Science 99

    Appendix 6: Detailed Program 119

    Appendix 7: List of Acronyms Used in the Report 131

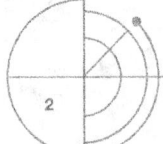

FOREWORD

From February 27 through March 2, 2007, the NASA Advisory Council Science Committee conducted the "Workshop on Science Associated with the Lunar Exploration Architecture" at the Fiesta Inn Resort in Tempe, Arizona. The workshop was planned and timed to feed into ongoing efforts by NASA's Lunar Architecture Team (LAT) to develop an exploration architecture for the return of humans to the Moon by 2020 in accordance with the Vision for Space Exploration (VSE) and the NASA Authorization Act of 2005.

The goals of the workshop were to: (1) ensure that NASA's exploration strategy, architecture, and hardware development enable the best and appropriately integrated science activities in association with the return of humans to the Moon and subsequent exploration of Mars; (2) bring diverse constituencies together to hear, discuss, and assess science activities and priorities for science enabled by the exploration architecture; and (3) identify needed science programs and technology developments.

The workshop was a key part of the Council's obligation to advise the NASA Administrator on science associated with the VSE while, in parallel, making its findings directly available to NASA's Exploration Systems Mission Directorate (ESMD) and Science Mission Directorate (SMD). The agenda was planned to cover exploration science, lunar science, lunar-based science, and science otherwise enabled by the emerging exploration architecture. Specific science objectives were discussed and priorities assessed as initial guidance for the planning of the return to the Moon program. The workshop deliberations and the ensuing assessments from the science subcommittees are intended to enable the Council to make recommendations to the Administrator relative to the exploration strategy and architecture being developed by NASA. The workshop served as a major venue for the science community to provide input through the Council and Science Subcommittee representatives. In addition to the Science Committee and Subcommittee representatives, approximately 75 topical experts made presentations and assisted with assessment of science objectives and priorities. An ad-hoc Outreach Committee was established to consider the key messages to be communicated to the science community and the public regarding the workshop outcome. The workshop was also open to the science community and the public. Some 250 attendees were present at the beginning of the workshop.

The workshop was organized by a committee consisting of representatives from each of the science subcommittees of the Council working with the subcommittee executive secretaries and representatives from SMD and ESMD at NASA Headquarters. The workshop was organized primarily according to subcommittee disciplines. Following an opening plenary with presentations by NASA officials and science community representatives to set the context, the workshop proceeded with breakout sessions designed to address the science objectives appropriate to each of the subcommittees as well as several cross-cutting themes. Subcommittees worked from objective lists developed by the LAT from April to December 2006. Assessment of priorities and recommendations stemming from that effort are detailed in this report and its appendices. Members of the LAT as well as program managers and others from NASA Headquarters, participated in each of the breakout sessions. Scientists considered potential constraints imposed by the exploration architecture and provided results of assessments directly to members of the LAT and to the workshop Synthesis Committee representatives. This report by the Synthesis Committee summarizes and formalizes those assessments for consideration by the Council's Science Committee and the full Council.

Logistics for the workshop were supported by SMD, ESMD, and the NASA Advisory Council. On behalf of the Science Organizing Committee and the participants of the workshop, gratitude is expressed to NASA for its support of this activity. Contributions from the many participants of the workshop are likewise gratefully acknowledged. As with the Falmouth Conference of 1965—which served to define and assess science objectives prior to the Apollo Program, and was both preceded and followed by other conferences—this report is not a final or complete document on the subject. Instead, it is the beginning of a fruitful partnership and process through which the science community can have input to NASA's exploration architecture and implementation of the Vision for Space Exploration.

Bradley L. Jolliff, Workshop General Chair

FOREWORD

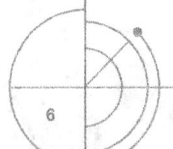

ORGANIZING COMMITTEE

Bradley L. Jolliff (Chair) — Council Science Committee, Washington University, St. Louis
Clive R. Neal (Vice Chair) — Lunar Exploration Analysis Group Chair, University of Notre Dame

Astrophysics
Heidi Hammel — Space Science Institute
John C. Mather — Goddard Space Flight Center (GSFC)
Eric P. Smith — NASA SMD, Executive Secretary

Earth Science
Michael Ramsey — University of Pittsburgh
Kamal Sarabandi — University of Michigan
Jean-Bernard Minster — University of California, San Diego
Lucia S. Tsaoussi — NASA SMD, Executive Secretary

Heliophysics
James Spann — Marshall Space Flight Center (MSFC)
David S. McKay — Johnson Space Center (JSC)
Barbara Giles — NASA SMD, Executive Secretary

Planetary Protection
Nancy Ann Budden — Office of Secretary of Defense
Andrew Steele — Carnegie Institute of Washington
Catharine A. Conley — NASA SMD, Executive Secretary
John D. Rummel — NASA Headquarters

Planetary Science
Charles K. Shearer — Curation and Analysis Planning Team for Extraterrestrial Materials (CAPTEM) Chair, University of New Mexico
Ariel D. Anbar — Arizona State University
Larry A. Taylor — University of Tennessee, Knoxville
Lars Borg — Lawrence Livermore National Laboratory
Michael H. New — NASA SMD, Executive Secretary

NASA Advisory Council
Harrison H. Schmitt — Chairman
Christopher Blackerby — Council Executive Director
Louis H. Ostrach — ESMD, Executive Secretary, Exploration Committee
Gregory J. Williams — SMD, Executive Secretary, Science Committee
Mark S. Robinson — Science Committee, Arizona State University
Gerald L. Kulcinski — Chair, Human Capital Committee, University of Wisconsin-Madison; Outreach Committee Chair for the Workshop

NASA Headquarters
Michael J. Wargo — NASA ESMD Lead and ESMD Lunar Exploration Scientist
Lisa D. May — NASA SMD Lead
Marc Allen — NASA SMD
Marian Norris — NASA SMD

The Workshop Synthesis Committee includes most of the members of the Organizing Committee listed above with the addition of Dr. Paul Hertz, NASA SMD.

ORGANIZING COMMITTEE

EXECUTIVE SUMMARY AND KEY ASSESSMENTS

The overall objective of the workshop was to provide input from the scientific community through the NASA Advisory Council to the NASA Administrator regarding science associated with the return-to-the-Moon phase of the VSE. Findings developed during the "Workshop on Science Associated with the Lunar Exploration Architecture" are intended to form a basis for Council recommendations regarding planning and implementation of NASA's Lunar Exploration Architecture and related science programs. Through attendance of their representatives, workshop considerations and findings became immediately available to the two Mission Directorates.

The workshop brought together diverse groups of space scientists with others who represented NASA SMD programs, science program managers, ESMD personnel, and LAT members involved in planning lunar exploration. Breakout sessions were organized mainly along science discipline lines and focused on:

- defining the key objectives of science associated with, or enabled by, lunar exploration;
- discussing implementation to achieve the objectives
- establishing overall science priorities within disciplines
- prioritizing objectives within the framework of the emerging lunar architecture, in particular those relevant to a polar outpost

The participants in the workshop had access to a variety of previously released studies in the context of their discussions. Studies that were referred to included, but were not limited to, the Report on the Scientific Context for the Exploration of the Moon (Space Studies Board), the Astrophysics Enabled by the Return to the Moon Workshop (Space Telescope Science Institute), the Earth Science Decadal Survey (Space Studies Board), and the lunar exploration white paper of the Field Exploration and Analysis Team (FEAT).

Findings and Key Assessments

Each of the subcommittees effectively identified the highest priority science activities that could be accomplished or enabled by the lunar exploration architecture and by the specific notional architecture that has been proposed, i.e., a lunar polar outpost. They also identified the highest priority science activities for a lunar outpost, as well as the general priorities for all science objectives for each subcommittee discipline. Some of the key findings are summarized in the following sections; however, the reader is encouraged to consider each of the discipline workshop reports (appendices 1–5) in their entirety. Only a brief summary is given here. A premise of the report (and of the workshop deliberations) is that the architecture will be based on an outpost concept. Assessments are made with respect to the notional polar outpost location, but attendees understood that a polar location and the sequence of missions to establish such an outpost are not set in stone at this time.

Astrophysics

The principal advantage offered by the Moon as an observatory platform for astrophysics is the radio-quiet environment of the lunar far side and the potential to place low frequency (meter-wave) radio telescopes there. Because of shielding from terrestrial (continuous) and solar (half-time) radio emissions and the lack of a lunar ionosphere, a far-side observatory offers the potential for extremely sensitive probes of the cosmic evolution of the universe. If 21-cm radiation from hydrogen emitted early during the formation of structure in the universe can be detected, this highly red-shifted 21-cm signal would provide a unique and sensitive probe of cosmic evolution, including the formation of the first structure in the universe and the first luminous objects. Appropriate steps must be taken throughout the architecture planning to ensure that a sufficiently radio-quiet environment can be maintained on the lunar far side, and that infrastructure is planned to enable eventual deployment of such a facility.

Some of the observatories envisioned by the astrophysics community for deployment beyond Earth would optimally be placed in free space. Future exploration program assets may enable or enhance deployment of such observatories. Others may be advantageously placed on the lunar surface, including low-frequency radio observatories, retroreflectors, and small competitively selected "payloads of opportunity." As the exploration program evolves, NASA should sponsor studies to investigate ways in which the exploration architecture can be enabling for astrophysics missions identified in science-community planning processes. The astrophysics community should consider these potential enabling capabilities as it conceives programs to yield the highest priority science within the available resources.

Earth Science

The Earth Science community is the one for which the lunar exploration architecture has the most critical implications owing to the potential siting of an outpost at a location that has a limiting view of Earth. Earth observations, which arguably have the most immediate societal relevance of all of NASA's science enterprises, require a vantage point from which the whole Earth is in constant or nearly constant view. The Earth-facing side of the Moon provides a unique observation point with the potential for full-disc and full-spectrum, long-term observation of Earth's changing surface and atmosphere from a stable and serviceable platform. If a polar outpost site were to be chosen, then to realize this potential, the architecture should include provisions for mobility to permit access to a suitable location such as the slopes of an Earth-facing massif such as Mount Malapert, near the lunar South Pole.

Given suitable viewing geometry, a dedicated Earth observatory on the Moon would allow for global, continuous full-spectrum views of the Earth to address a range of Earth science issues. Such issues include detection and analysis of time-dependent atmospheric composition (global mapping of emissions, long-range transport of pollution plumes, greenhouse gases sources and sinks), ecosystem monitoring, observation of changes in the polar caps and other aspects of the cryosphere, and vertical structure of the Earth's atmosphere. A lunar platform also allows the Sun-Earth system to be observed simultaneously, providing data on the Earth's radiation balance and the influence of solar variability on climate. An observatory sited on the lunar surface would allow for growth and serviceability over time. A facility at the lunar outpost could also serve as a communications bridge across satellite platforms in other orbits (e.g., low-Earth, geostationary, Global Positioning System [GPS]), providing for synergistic operation and fusion of results of Earth-observing assets.

Heliophysics

Heliophysics science directly addresses one of the major hazards of space-faring explorers, i.e., the solar radiation environment. Therefore, many of the heliophysics observations and science objectives are enabling for exploration in that they are necessary for safe and sustained operations on the Moon. The Moon is also seen as a historical record of past solar activity in much the same way as ancient ice cores on Earth provide a record of atmospheric composition and activity, as well as overall climate change. Heliophysics shares with Earth science the objective of studying and monitoring interactions of the Sun and its fields with Earth's atmosphere and fields. Such observations have great impact for society in terms of anticipating short-range hazards to space and satellite operations, as well as understanding long-term responses of Earth's fields and atmosphere to the Sun's radiation and plasma environment.

Of the listed LAT objectives for heliophysics, 13 are considered to be high-priority science. Five of these relate to understanding the lunar environment, and several should be completed prior to sustained human activity. Of special concern is researching the electromagnetic and charged dust environment at a potential lunar outpost or other sites. Several of the high-priority objectives are best accomplished from orbit and could be deployed as "drop-off" satellites. Real-time space weather monitoring would not be done on the Moon; monitoring measurements must be located as close to the solar source as is feasible. Lunar surface operations should include, as a high priority, sampling of the surface regolith and regolith strata within the context of well understood stratigraphy—such as within volcanic flow sequences—in order to investigate the nature and history of solar emissions and galactic cosmic rays. Although trenching could be done at a polar outpost site, access to a surface with inter-layered regolith and ancient datable lava flows would be particularly significant. Improved measurements of the solar wind composition and the flux and composition of interplanetary and interstellar grains bombarding the lunar surface, and imaging of high energy x rays and gamma rays, can also be accomplished on the lunar surface.

Planetary Science

Planetary science seeks to understand the origin of the solar system, the diversity of its planets and moons, and the factors involved in the origin and sustainability of life on Earth and perhaps on other planets and moons. For planetary science, the Moon is the keystone recorder of early solar system processes, especially those pertaining to the Earth-Moon system and other terrestrial planets. The most important processes on the early Earth that shaped the environment in which life originated are recorded on the surface of the Moon in a uniquely accessible way.

A polar outpost will provide access to one of the major unexplored and unsampled regions of the Moon (i.e., the poles), adding significantly to knowledge gained through exploration of the central near side by Apollo. A southern polar site also would be on the rim, potentially providing access to associated materials of the huge and ancient South Pole-Aitken Basin. As the earliest and largest impact basin that has been identified on the Moon, it is also a key to deciphering the early heavy bombardment history of the Moon and Earth. Many other planetary science objectives can be addressed at an outpost, but most require access to multiple locations around the entire Moon. Mobility—both short-range near the outpost and long-range away from the outpost—is highlighted as a key asset to accomplish high-priority science. Robotic missions, both before and during human exploration of the Moon, are intended to accomplish some of the highest priority science, primarily to access sites on the Moon far distant from the outpost site. Planetary science holds among its highest priorities the development of technologies and strategies to deploy and maintain a geophysical network for a long duration, with broader implications for planetary exploration in general.

As was the case with Apollo, one of the anticipated results of the return of humans to the Moon is the acquisition of carefully selected, collected, and documented geologic sample materials—rocks and regolith—for study in terrestrial laboratories. Strategies, technologies, and operational techniques to maximize the mass and vertical and lateral diversity of returned lunar samples, as well as detailed documentation of the locations and associations of these samples as developed during Apollo, must be modernized in the form of new protocols. Crew composition, crew training, and documentation efficiency are critical to the successful return of samples. As the lunar exploration architecture develops, plans for geological and geophysical field training will be an essential component in the preparation of astronaut crews for future missions to the Moon. The Orion Crew Exploration Vehicle (CEV) should have a capability similar to the Apollo science instrument module (SIM) to facilitate scientific measurements and the deployment of payloads from lunar orbit—a capability also important to heliophysics. Lastly, continuous scientific input should be an integral component of the decision-making process for landing-site targets for a lunar outpost or any lunar mission.

Planetary Protection
Operations on the Moon are not constrained by current planetary protection restrictions. This makes the Moon an optimal location to establish the magnitude of forward contamination associated with human exploration. The lunar return can facilitate development and testing of equipment and technologies designed to limit human-associated contamination.

Contamination-control technologies for planetary protection must be developed before human missions to Mars can occur. Tests such as the prevention of back contamination of sample containers by the extremely fine-grained lunar dust can serve as a surrogate objective for preventing such contamination of sample containers planned for use on Mars. Technologies and experimental equipment to perform planetary protection assays will need to be included in up-mass to the lunar science laboratory, and crews will need to be trained in operation of equipment.

The lunar environment may also have some aspects needing protection (e.g., polar volatiles). Planetary protection science objectives with high priority include: future in situ investigations of locations on the Moon by highly sensitive instruments designed to search for biologically derived or other organic compounds, which will provide valuable "ground truth" data on in situ contamination of samples; study of lunar spacesuit competency, containment, and leakage; the ability of evolving suit requirements to affect Mars suits, habitat designs, and requirements; and understanding possible contamination of lunar ices with non-organically clean spacecraft.

Cross Cutting Issues
Trade studies should consider options for outpost and observatory siting. This issue is especially important for Earth observations, but also plays a role in the objectives for heliophysics and planetary science. Clearly, access to solar power has contributed to the decision to select a polar location for the notional architecture; however, options and trade studies for outpost locations at other latitudes are needed.
The roles and capabilities of astronauts in the deployment, operation, and servicing of science activities, as well as in regard to sampling, instruments, and facilities within the context of the planned architecture, need to be clearly defined and supported. Much experience has been gained through Apollo, Skylab, the Shuttle, Hubble servicing, and the International Space Station (ISS). For full mission success, many of the science objectives will necessarily require the involvement of a scientist-astronaut serving as an integral part of the science experiment or as a field geologist.

Science activities enabled by lunar exploration should continue to be evaluated and prioritized by the decadal survey and science roadmapping processes. Lunar science assessments formulated at the workshop are not intended to supersede the decadal survey process, but should be considered as input to the next National Research Council (NRC) decadal surveys and NASA science roadmaps. SMD has a well-validated process for establishing science priorities within their resource allocations. Once complete, information pertaining to the lunar science opportunities should enter into this process in the same manner as other SMD pre-planning activities.

Implementation of the VSE should be planned to accommodate capabilities that will enable the highest priority science, at least to the extent that other major objectives are not compromised. Because science missions are competed and not set forth in a specific programmatic way, the exploration architecture should be designed to enable the kinds of activities that are listed as being of potentially high scientific priority, even though some of these activities may never actually be done. This approach proved to be highly advantageous and flexible during Apollo.

Regular reviews (e.g., through the NASA Advisory Council structure) of major LAT decisions that may influence the science productivity of the lunar architecture should be conducted. Future evaluations of science objectives must assess the cost effectiveness of lunar outpost implementations versus implementations that utilize robotic/unmanned missions around the Moon or elsewhere.

Outreach

The Outreach Committee was integrated with subcommittee breakout groups throughout most of the workshop, but convened toward the end of the workshop to articulate the main messages of the workshop to the scientific community and to the public. These are as follows:

1. The Moon is witness to 4.5 billion years of solar system history. Human exploration of the Moon will contribute greatly to discovering the origins of the Earth and of humanity.
2. The Moon is a unique location from which to observe and analyze the ever-changing nature of the Earth, Sun, and universe.
3. The Moon is a fundamental stepping stone to the human exploration of Mars and the rest of the solar system.

The Outreach Committee also formulated messages relative to each of the subcommittee disciplines. These messages are listed in the body of the main report.

Concluding Statement

As with any new phase of space exploration, the scientific possibilities associated with the return-to-the-Moon exploration architecture are numerous and exciting. In this case, the human element brings a unique set of capabilities, and the global exploration strategy associated with the VSE offers the potential to extend the possibilities considerably. The intent of the Tempe workshop was to provide a clear assessment of the science priorities for activities enabled by the exploration plans and architecture. At the workshop, the science community began this process, making substantial progress especially through interactions between individuals and groups that represent U.S. stakeholders in space science and exploration. We hope to continue this process as development of the exploration architecture progresses in coming years, leading to the return of human space flight to the Moon and preparation for the journey beyond.

EXECUTIVE SUMMARY AND KEY ASSESSMENTS

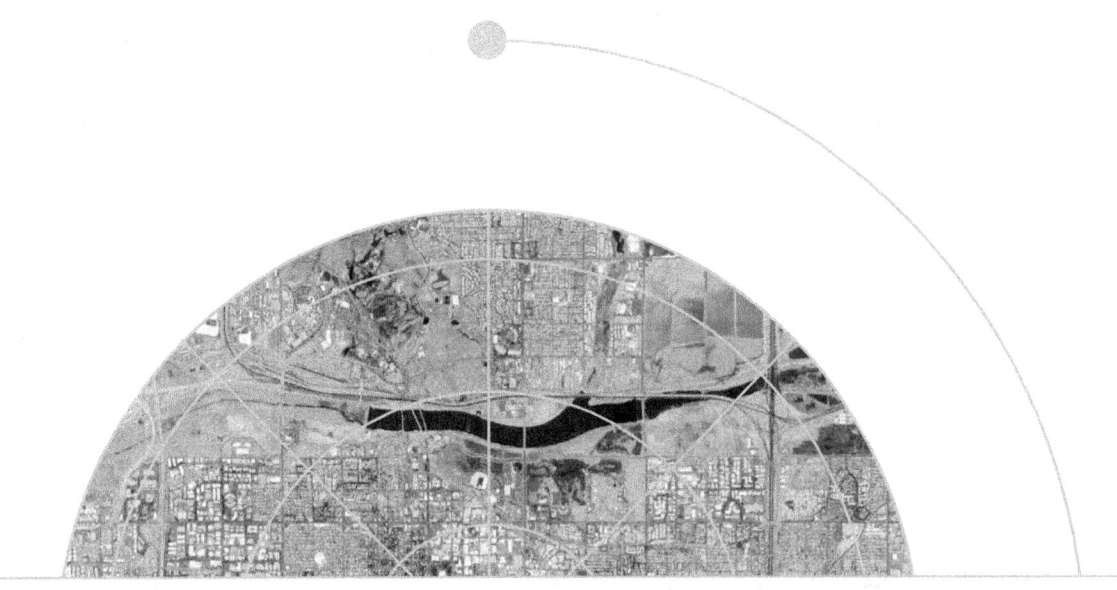

INTRODUCTION AND OVERVIEW
OF THE WORKSHOP

The NASA Advisory Council Workshop on Science Associated with the Lunar Exploration Architecture was held February 27 through March 2, 2007, at the Fiesta Inn Resort in Tempe, Arizona. The workshop was planned to bring together the science subcommittees representing the various space science disciplines within NASA, representatives from SMD, ESMD, and LAT, with members of the space science community at large to discuss science activities associated with or enabled by the emerging exploration architecture. The workshop is part of an ongoing effort to advise NASA on the exploration architecture associated with the VSE.

One of the goals of the workshop was to bring together the diverse constituencies for science activities associated with the VSE and the return of humans to the Moon. The first day of the workshop provided an overview of the activities and science objectives being considered as part of this new exploration program so that all participants could gain a sense of the diverse potential activities and begin the process of assessing the various activities in terms of priorities, architecture capabilities and requirements, and phasing. The workshop opened with a plenary session during which NASA officials, including the Administrator, Dr. Michael Griffin, laid out plans for implementing the VSE. These plans include NASA's Global Exploration Strategy and specific efforts to delineate a notional lunar exploration architecture. Details of the notional architecture are given in a subsequent section as context for the subcommittee assessments of science activities and objectives. Development of the exploration architecture and constraints imposed by the architecture are not necessarily—or even mostly—driven by science considerations. The key issue for the science community was to determine the highest priority science activities that could be accomplished within the constraints of the exploration architecture. Of course, wherever possible and wherever warranted, the science community should (and did) suggest changes to the architecture that would accommodate or enable high-priority science activities without compromising other key objectives.

Lunar science priorities are also under study by the NRC, and results from the interim report on the Scientific Context for Exploration of the Moon were presented during the opening plenary session. The NRC's recently released Earth Science Decadal Survey was also available; however, this decadal survey does not address Earth science possibilities enabled by lunar exploration. Each of the subcommittees, including Astrophysics, Earth Science, Heliophysics, Planetary Protection, and Planetary Science, presented an overview of the topics to be addressed during the workshop. The full agenda is included as Appendix 6 of this document.

One of the specific goals of the workshop was to have the science community—through the subcommittees, invited experts, and other participants—provide assessments and prioritization of science objectives that had been previously identified and grouped according to discipline by the LAT. The process of identifying potential objectives began in April 2006 at ESMD's Lunar Exploration Workshop and continued through December 2006 with a series of reviews and assessments by various organizations, including the Council subcommittees, the NRC lunar science study, and various special action teams organized by the Lunar Exploration Analysis Group (LEAG). Independently, FEAT—organized through the efforts of the University of Texas, the University of Wyoming, and Arizona State University—provided attendees with a detailed lunar field exploration white paper and sponsored a pre-workshop geological-methods field trip. At the Tempe workshop, the subcommittees provided assessments to prioritization, phasing, needed technology developments, and other issues connected to the final list of science objectives. These important assessments are included in this report as Appendices 1-5. The final assessments of each of the subcommittees were briefed at the closing plenary session on March 2.

A key element of the workshop was the activity of the Outreach Committee. This committee was charged with determining how best to communicate the results of the workshop to the broader science community and the public. The committee, which included members from each of the science subcommittees as well as outreach specialists, prepared a set of high-level highlights for science activities discussed by each of the subcommittees. These highlights are detailed in a subsequent section.

The workshop was purposely designed as a meeting open to any interested persons to ensure the input of the science community in the development of the exploration architecture. The plenary sessions of the workshop were broadcast over the Internet using WebEx technology. Finally, in addition to this report, individual presentations made during the workshop and white papers submitted to the workshop for oral or poster presentation or print-only have been placed on a public access Web site hosted by the Lunar and Planetary Institute, Houston, Texas, at *http://www.lpi.usra.edu/meetings/LEA/*.

INTRODUCTION AND OVERVIEW OF THE WORKSHOP

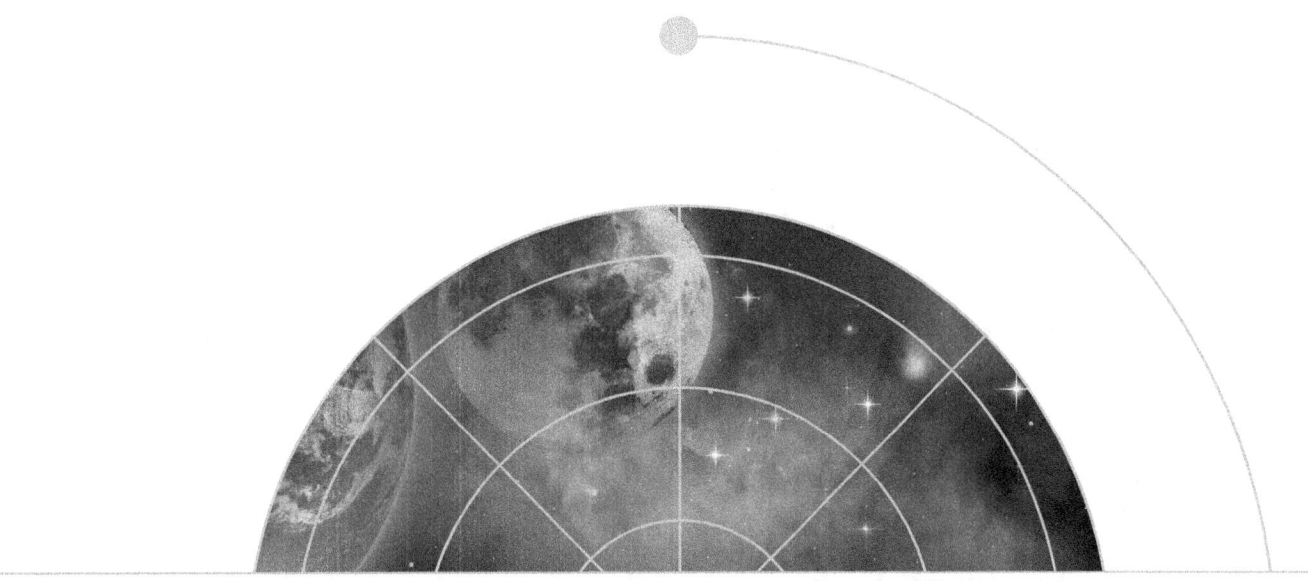

THE VISION FOR SPACE EXPLORATION
AND THE ROLE OF SCIENCE

The Vision for Space Exploration as set forth by the President in 2004, and the role of science within the VSE, were articulated in the opening plenary by Dr. Michael Griffin and by Deputy Associate Administrator for Science, Dr. Colleen Hartman. The mandate of both the President and Congress to extend NASA's human exploration beyond low-Earth orbit (LEO), beginning with the establishment of a sustained presence on the Moon, was expressed very clearly.

Such exploration endeavors, however, require multinational efforts if they are to be affordable and of maximum global impact. This requirement forms a basis for the lunar-outpost approach for this next phase of human exploration. Specifically, NASA plans to provide the key infrastructure elements and core capabilities, including transportation and communication systems, but will develop international and commercial partnerships to augment other aspects of the exploration program, such as these of mobility, habitats, robotic capabilities, and science activities. Current budget levels and projections may constrain some important elements such as robotic precursor missions; however, such areas are ripe for augmentation of core capabilities by international partners or through science-focus missions by means of established programs. If critical components of the architecture (e.g., mobility and habitats) are not forthcoming from international sources, NASA will seek the funding needed to provide them.

The model that has been described for the lunar outpost concept has a useful analog in the historical exploration of Antarctica and the establishment of research outposts such as the Base at McMurdo. Legal aspects of lunar activity would be constrained by the Outer Space Treaty of 1967, portions of which were influenced by the Antarctic Treaty Regime. In such an endeavor, the role of science, while not the only factor, is key. Not only are good science opportunities enabled by this architecture, but science is also enabling for exploration. Heliophysics, for example, provides an understanding of and predictive capabilities for space weather, including potentially deadly solar radiation events. Planetary science includes the identification of hazards, as well as the delineation of lunar resources and approaches to utilize them.

> *"Science in the space exploration vision is both enabling and enabled."*
> —*President's Commission on Implementation of U.S. Space Exploration Policy*

An expressed concern among some scientists across disciplines relates to the role of science activities enabled by the human exploration program and identified as compelling by the science community. Specifically, will such science activities have a greater or lesser priority than other space science activities previously planned? NASA clearly stated that the Agency will continue to use community-based processes such as decadal surveys and science roadmaps, as well as the Council's activities, to ensure that science investigations enabled by exploration are effective, relevant, and of the highest scientific quality.

The Moon is a stepping stone for space exploration—a proving ground for the eventual expansion of human activities to Mars. Humans must learn to live and work off Earth and beyond low-Earth orbit. Humans will first accomplish these goals on the Moon.

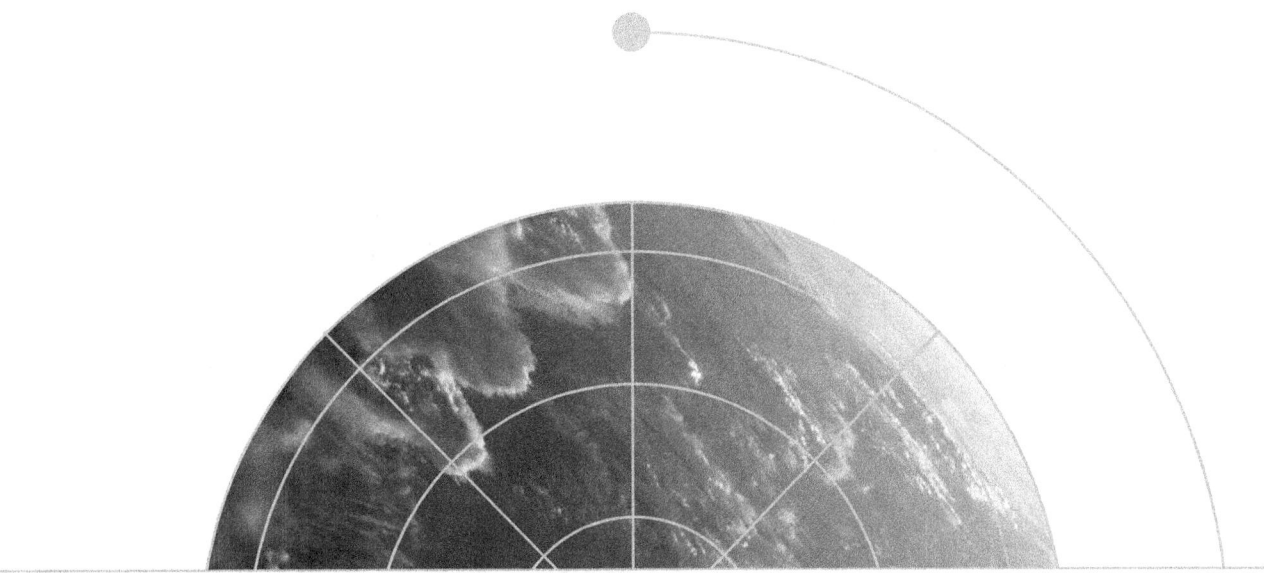

THE GLOBAL EXPLORATION STRATEGY

In April 2006, NASA held a workshop to begin the process of defining exploration objectives across a wide range of themes. Attendees came from the science community, commercial sector, and international agencies, including all of the International Space Station partners. From this workshop, a strategy was developed to respond to two key issues: why we are going back to the Moon, and what we hope to accomplish when we get there.

Six overarching exploration themes were identified in response to the first issue as follows:
- Human civilization
- Scientific knowledge
- Exploration preparation
- Global partnerships
- Economic expansion
- Public engagement

In regard to the second issue, 180 specific lunar exploration objectives were defined that fall within the six exploration themes. Over 1,000 people from around the globe, including experts from 14 space agencies, commented on and contributed to these themes and objectives. This global approach allows for the inclusion of a diverse group of stakeholders—international partners, academia, industrial and commercial sectors, and private U.S. citizens—in the strategy development process. As a result, a strong foundation for further discussion and cooperation has been developed in recognition of the fact that the next phase of human exploration must have broad support, including international partnerships. In the long term, these exploration goals include the Moon, Mars, and destinations beyond Mars; however, the initial focus is on human exploration of the Moon.

As part of developing the global exploration strategy, NASA identified 40 objectives of particular interest. These objectives include activities associated with preparing for human missions to Mars and other destinations; providing the capabilities to support scientific investigations; activities that would enable an extended/sustained human presence on the Moon, such as demonstrating the use of in situ resources and measuring lunar phenomena, analyzing lunar resources, and characterizing their possible use; activities that would enable international participation; and activities that would engage, inspire, and help educate the public.

Moving from an exploration strategy to an exploration design required development of a notional architecture that takes into account budget and technology constraints and that achieves national goals while maximizing response to the six exploration themes. Such a notional architecture permits assessment of how well the specific objectives might be met. From consideration of the broad set of themes and the specific objectives, NASA determined that a lunar outpost rather than a set of sortie missions would enable a sustained human presence on the Moon that meets the priorities of the VSE.

Although the lunar exploration architecture includes a wide variety of activities—ranging from the transportation and communication infrastructure to habitat development and on-surface exploration activities—a key aspect to developing the architecture is identifying and differentiating those areas in which NASA would take the lead and those areas that would have a primary science focus or would be well suited to commercial or international involvement. Key areas for NASA development include, but are not limited to, space transportation (including the Orion CEV, the Ares I and Ares V rockets, and the Lunar Surface Access Module), initial communications and navigation capabilities, the development of a spacesuit for extravehicular activity (EVA) on the lunar surface, providing a closed-loop life support system, and obtaining knowledge about the effects of the lunar environment on humans. Development of habitat elements and surface activities, including in situ resource utilization, scientific experiments, and on-surface mobility, are examples of areas in which the involvement of the commercial, international, and scientific communities could augment the exploration infrastructure.

NASA's approach to sustaining a human presence on the Moon is based on a "go-as-we-can-afford-to-pay" approach that would enable humans to return to the Moon no later than 2020. Through the combined exploration activities, NASA intends to extend operational experience in the hostile extraterrestrial planetary environment and to develop experiments and demonstrations to characterize the planetary environment. Planned operations at the lunar outpost include demonstration of the feasibility of in situ resource utilization (ISRU), and will provide opportunities for scientific investigation, economic expansion, education, and international participation—all of which will help to prepare the way for the human exploration of Mars.

THE GLOBAL EXPLORATION STRATEGY

THE LUNAR EXPLORATION ARCHITECTURE

The notional lunar architecture presented at the Tempe workshop was that of the south polar outpost, as first described at the December 2006 Exploration Conference held in Houston, Texas. This architecture follows from NASA's human exploration objectives and national priorities coupled with the global exploration strategy described in the previous section.

The polar outpost scenario follows from consideration of five key questions:

1. What are the U.S. priorities and phasing for what will be achieved at the Moon?
2. How do priorities drive important decisions? Key decisions include whether to design to an outpost or to engage in a series of sortie missions, where to locate the landing sites, and how much flexibility will be needed to address other U.S. priorities or far-term interests.
3. What infrastructure is required to support priorities? Considerations include schedule vs. flight rate and cost vs. available budget.
4. What will NASA plan on developing (i.e., critical-path hardware to achieve primary objectives) while allowing for parallel developments from commercial and/or international communities?
5. What level of limiting resources will allow for optimum realizable capability? Such resources are enabled by the basic NASA transportation architecture, including down-mass and up-mass at the Moon and power generation at the outpost.

Consideration of three of the exploration themes drove the architecture development to an outpost. These are Exploration Preparation, Human Civilization, and Economic Expansion. The outpost concept is thought to better enable global partnerships and allow development and maturation of ISRU efforts. According to LAT assessments, the outpost concept should result in the quickest path toward other destinations. Moreover, many science objectives can be satisfied at an outpost.

The decision to select a polar location for the notional outpost site (see figure 1) stems from several lines of argument. In terms of safety, a polar location provides good opportunities to return and the ability to abort to surface from orbit. This location also permits a relatively low-energy return to Earth, and its high percentage of sunlight allows the use of solar power while shortening the time required to sustain outpost operations through the lunar night. Initially, this access to plentiful power potentially makes a polar site more cost effective than one at lower latitudes that would require other sources of power. Although oxygen is as abundant in the soil there, as it is everywhere on the Moon, the polar site permits access to regolith with enhanced hydrogen concentrations and possibly water-ice (as well as other needed volatile elements) in permanently shadowed craters. In general, a polar site offers the flexibility of an incremental buildup using solar power, enhanced surface daylight operations, one communication asset (with backup), and extended opportunities to launch. As a location that is far removed from previous lunar landing sites and well known areas, a polar site offers an element of exploration excitement and a unique environment of proximity to cold, dark craters.

Clearly, a key driver of the polar locations is access to solar power. The south polar site at the rim of Shackleton crater (see figure 2) has a zone of 70 percent illumination during the middle of southern winter and better during southern summer. This site is within several kilometers of areas of permanent shadow within Shackleton crater. The Moon's north pole has three areas that experience 100 percent sunlight during northern summer and two zones that are proximal to craters in permanent shadow. Detailed mapping and imaging by the polar-orbiting Lunar Reconnaissance Orbiter (LRO), scheduled for launch in 2008, will better define the areas at the poles that are subject to constant or near-constant illumination.

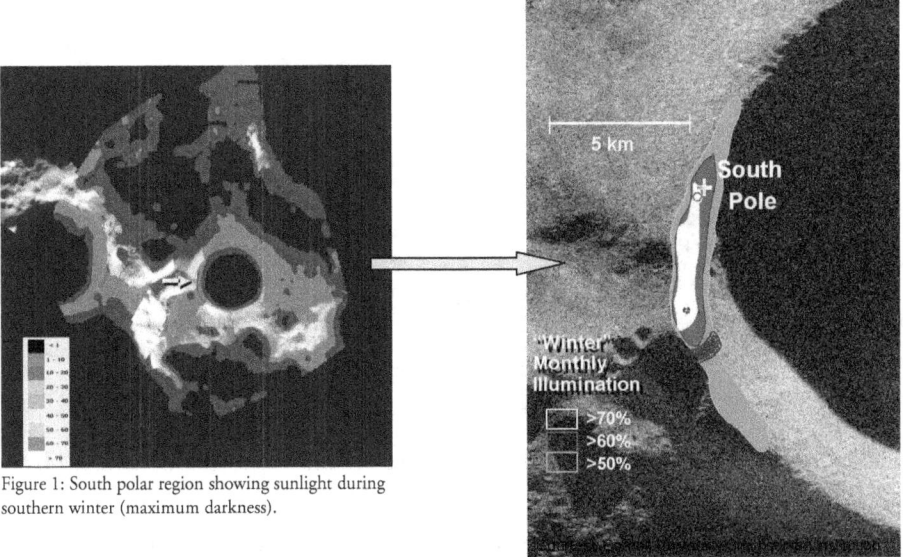

Figure 1: South polar region showing sunlight during southern winter (maximum darkness).

Figure 2: Possible location of an outpost on the rim of Shackleton Crater.

Although a polar site is considered for the notional architecture, the transportation infrastructure and the lunar lander design are intended to enable human sortie missions to any location on the Moon, including the top ten sites identified by the 2005 Exploration System Architecture Study (ESAS) Report.* Options exist within the architecture to trade crew for science payloads, including sample-return mass. For the outpost architecture, the point-of-departure design envisages landing missions beginning in 2020 at 6-month intervals that will incrementally build up outpost infrastructure and capabilities, including habitation modules, solar power collection and storage units, surface mobility and other logistics equipment, and ISRU modules. The ability to fly human sorties and cargo missions with the human lander will be preserved. The initial power architecture will be solar with the potential augmentation of nuclear power at a later time. Robotic missions will be used to characterize critical environmental parameters and lunar resources, as well as to test technical capabilities as needed. The ability to fly robotic missions from the outpost or from Earth will be a possible augmentation.

NASA's implementation philosophy is that the U.S. will build the transportation infrastructure, the initial communication and navigation systems, and initial surface mobility capability. The architecture is open; NASA will welcome external development or augmentation of lunar surface infrastructure. The U.S. will perform early demonstrations to encourage subsequent development and will welcome external cooperation and parallel development of capabilities initially developed by NASA.

Desired capabilities at the outpost include a mature transportation system, a closed-loop habitat, long duration human missions beyond LEO, surface extra-vehicular activity (EVA) and mobility, autonomous operations, advanced robotic missions, minimal reliance on Earth via in situ fabrication and resource utilization, and enhanced commercial and international partnerships.

---

* NASA's Exploration System Architecture Study, Final Report, NASA TM-2005-214062, November 2005.

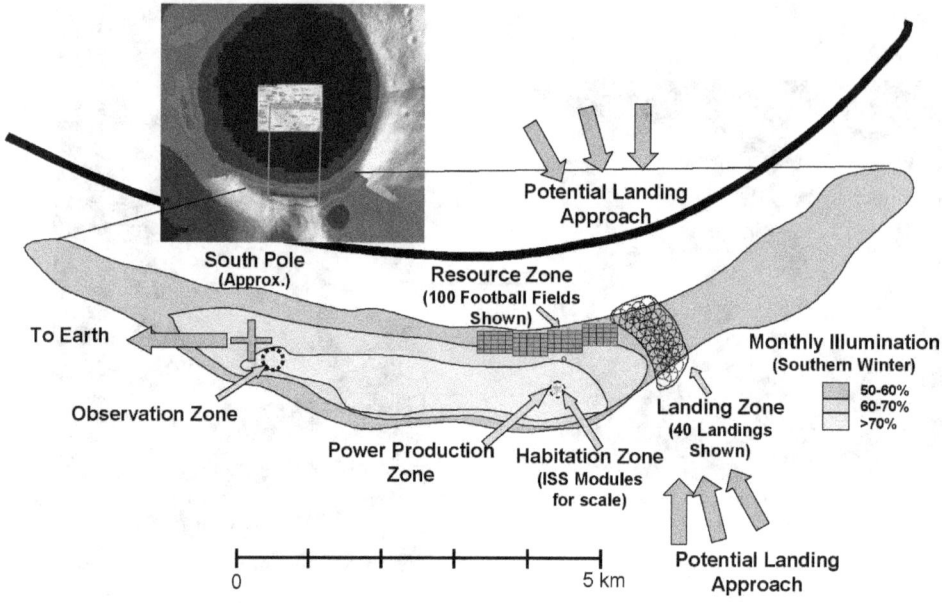

Figure 3: Notional activity zones at an outpost site located on the rim of Shackleton crater at the South Pole.

The architecture envisages follow-on to outpost construction that will possibly include human exploration of other lunar sites via sorties, expanded lunar outpost site operations, and expanded lunar outpost activities through commercial and/or international partnerships. Some of the exploration objectives for the outpost include the development and validation of tools, technologies, and systems that excavate lunar material; characterizing radiation bombardment on the lunar surface and subsurface; understanding the effects of the integrated lunar environment on human performance; and providing position, navigation, and timing (PNT) capabilities to support lunar operations, eventually evolving to support operations at Mars.

The lunar exploration architecture as articulated at the Tempe workshop was intended to be viewed as a point of departure. Subsequent activities of the lunar exploration architecture development include updating the objectives that drive the architecture, coordinating lunar exploration plans with interested communities, finding opportunities to collaborate with partners, refining campaign and architecture concepts, and refining hardware concepts for the different elements of the architecture. Following the initial phase of lunar architecture development, a similar effort will be undertaken to develop a Mars reference mission. The expressed intent throughout this process is to continue to engage academia, the private sector, and other stakeholders in defining a sustainable program of exploration.

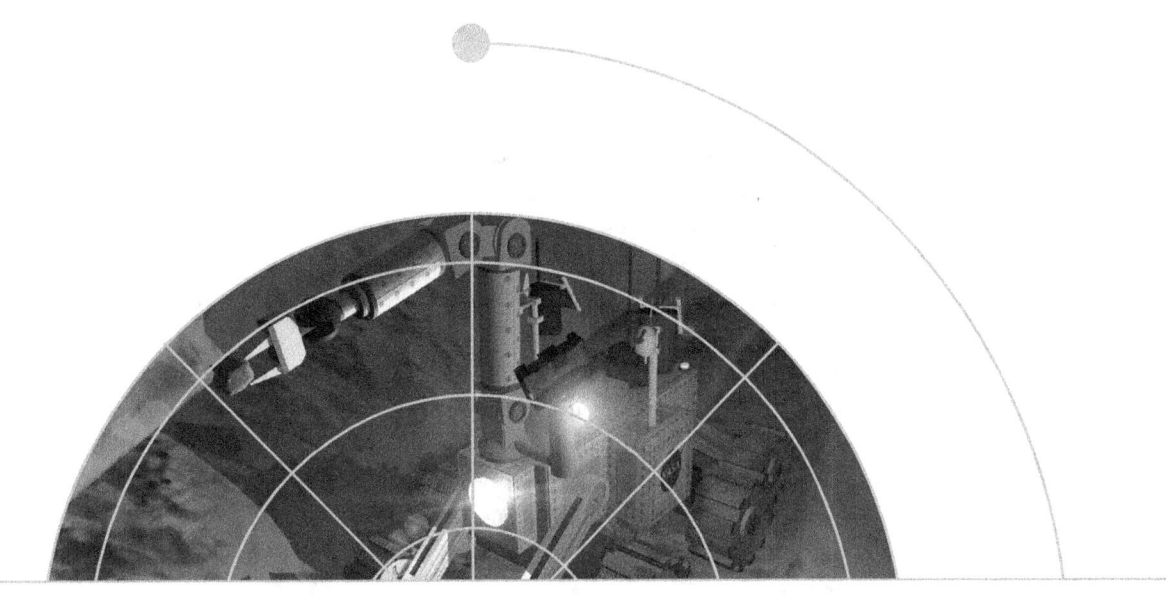

ASSESSMENT OF
POTENTIAL SCIENCE ACTIVITIES

Part of the development of the lunar architecture involved an assessment of science activities, conducted by the Science Capability Focus Element (SCFE) of the LAT. The results of this assessment conducted in 2006 were briefed at the Tempe workshop by Dr. Laurie Leshin, Director of Sciences and Exploration at the Goddard Space Flight Center and co-lead of the SCFE.

This assessment involved a distillation of all the science-related objectives into a set of 45 objectives that fall within the disciplines represented by the main divisions of the SMD. Each objective was analyzed to define needed capabilities and time-phasing issues and mapped to the lunar exploration architecture for "goodness of fit." The resulting assessment showed what scientific objectives could be relatively easily accommodated, as well as what changes might be needed to accomplish additional scientific objectives.

The matrix of science objectives was a focal point of the Tempe workshop, and participants were asked to evaluate the objectives in terms of science priorities and how these would map to the lunar exploration architecture. Each of the science subcommittees had previously seen and commented on the objectives, but the workshop was specifically tasked to present and discuss the related science issues and provide assessments of priorities for the potential science activities associated with these objectives.

Assumptions pertaining to capabilities for this assessment included the following:
- A polar outpost-based architecture—all missions in this phase would go to (or near) the outpost site (except for any orbital capability)
- A 4-member crew on initial 7- to 14-day stays followed by longer missions
- Some capability to fly robotic missions, especially before humans arrive
- Some moderate mobility (~10-20 kilometers) for the crew from the outpost site during the short-stay missions
- ~500 kg of payload down-mass for science experiments/tools on crewed missions
- ~100 kg sample return capability on crewed missions, including sample containers

In the original assessment done by the SCFE, the following rating criteria were used:
- [1] Objective can be substantially accomplished by 2025 within the current architecture, assuming the priority and funding are allocated.
- [2] Objective will very likely take longer than the 2025 time horizon to accomplish, but could be accomplished in an outpost-based architecture.
- [3] Some part of the objective can be accomplished within the current architecture by 2025.
- [4] Objective can be accomplished with a combination of outpost-based science and robotic sorties.
- [5] Objective can really only be accomplished through the addition of human sorties, selection of a different site for the outpost, or the addition of some other capability, such as long-range mobility, to the current architecture.

The initial overall assessment by the SCFE was positive, and it indicated that a polar outpost site would accommodate a large number of the science objectives, with over 50 percent potentially falling within the green rating. Science priorities for the objectives were not factored into the initial assessment; all objectives were treated with equal rank. Providing the priorities for specific objectives was requested of the Council and its science subcommittees to be accomplished at the Tempe workshop. The findings of the workshop (next section) reflect the efforts of the workshop participants to consider the listed science activities, evaluate priorities, and assess the implementation of activities to address the objectives within the notional architecture.

Workshop participants recognized that the notional architecture was intended to be a point of departure, and that results of the workshop, along with other inputs such as the NRC study on the Scientific Context for Exploration of the Moon, would be used to further refine or revise the notional architecture and to help determine the optimal time phasing and relationships of the various possible activities. The assessments and findings of the science subcommittees are summarized in the next section and placed within the overall context of the exploration architecture as presented at the workshop. These assessments include discussion of how the potential science activities might or might not fit within the exploration architecture, such as how some of the activities might be enabled by the transportation infrastructure (e.g., the Constellation Program), as well as outpost-specific issues. The full reports of findings and assessments are provided as appendices to this report.

ASSESSMENT OF POTENTIAL SCIENCE ACTIVITIES

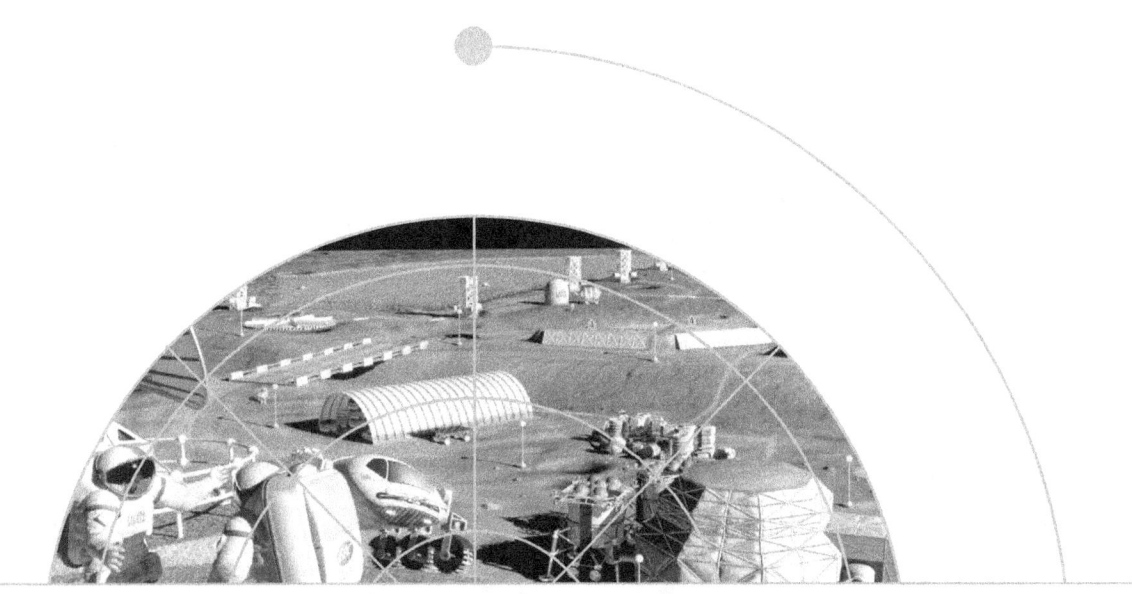

FINDINGS OF THE WORKSHOP

Astrophysics

Four science goals identified in the workshop encompass the breadth of current astrophysics research and are widely believed to pose intriguing astrophysical challenges for the next two decades.

These are:
1. What is the nature of the dark energy causing the cosmic expansion to accelerate?
2. Are there habitable extrasolar planets and, in particular, is there extraterrestrial life?
3. Which astronomical objects and which physical processes were involved in the "first light" and the reionization of the universe?
4. How did galaxies and the large-scale structure of the cosmic web form? The participants in the Tempe workshop agreed with these scientific goals and adopted them as a framework within which to evaluate the objectives crafted by the LAT.

Astrophysical Opportunities within the Lunar Architecture Worth Pursuing

Meter-wavelength radio telescopes on the lunar far side would have exciting applications in cosmology, extra-solar planet characterization, and the physics in the nuclei of active galaxies. Concepts for such telescopes have a reasonable science and technology expansibility from small precursors to eventually large facilities. Also, in that field, good synergies exist between heliophysics and astrophysics. A similar access to the lunar near side would be desirable for deployment of a widely dispersed retroreflector/transponder network to obtain increased accuracy for tests of general relativity. Smaller "payloads of opportunity" can provide interesting and competitive science. These smaller payloads, which should be competitively selected, do not necessarily conduct science of the highest (decadal survey) priority, but they still do solid science that meshes well with the lunar architecture.

Enabling Capabilities for Astrophysics

Radio-quiet (free from radio frequency interference [RFI]) environment and infrastructure on the lunar far side or near the Shackleton site for a meter-wavelength radio observatory. Because of its continuous shielding from terrestrial radio emissions and part-time shielding of solar radio emissions and lack of a lunar ionosphere, the far side of the Moon offers the potential for extremely sensitive probes of the cosmic evolution of the universe. During at least a portion of the formation of structure in the universe, the dominant baryonic component—hydrogen—should have emitted 21-cm radiation. If this radiation can be detected, this (highly) red-shifted 21-cm signal will provide a unique and sensitive probe of cosmic evolution, including the formation of the first structure in the universe and the first luminous objects. The most sensitive observations of these red-shifted 21-cm hydrogen signals will be obtained in a location on the lunar far side that is shielded from interfering signals.

Large launch vehicle capabilities. The Ares launch system offers a capability that could revolutionize astrophysics and other sciences by enabling entirely new classes of missions that will achieve priority observations. Current estimates for the launch mass and faring volume could enable:

- a 6- to 8-meter class monolithic UV/Visible/IR observatory
- a 5-meter cube (130,000 kilogram) gamma ray water calorimeter
- a 4-meter-class x-ray observatory
- a 15- to 20-meter-class far-IR/sub-mm observatory
- a 25- to 30-meter-class segmented UV/Visible/IR observatory
- a 150-meter-class radio/microwave/terahertz antenna
- constellations of formation-flying spacecraft

Capability for secondary payload of small or medium science instruments. The architecture should include the capability for secondary payloads on both the Ares launch vehicles and the Orion space vehicles. These capabilities could include features such as an Evolved Expendable Launch Vehicle Secondary Payload Adapter (ESPA) ring on the launch vehicles that could carry secondary payloads for deployment in near-lunar space. Also, the ESPA ring could form the structure for a secondary spacecraft like LCROSS (Lunar CRater Observation and Sensing Satellite) that would be deployed after the primary payload has been separated. Capabilities might also include secondary payloads for on-spacecraft autonomous instruments that do not require deployment. The Orion service module should also have the ability to carry secondary payloads that could be deployed in lunar orbit, and have a payload bay that could accommodate remote sensing and in situ experiments with the necessary thermal, mechanical, power, and data handling interface.

In-space operations. Very large aperture systems and spatial interferometers will be necessary to achieve many of the highest-priority astrophysics goals. Such systems must operate at various locations in free space throughout the Earth-Moon system, such as libration points, and high-Earth and geosynchronous orbits. Capabilities to support these high-value systems will eventually become essential (e.g., assembly, service, repair, refuel). Such capabilities may be achieved by augmentations to NASA's Exploration Architecture, which will be operational during the same timeframe. Examples of enabling capabilities include robotic or telerobotic systems, advanced in-space EVA from Orion, and capable transportation (including the Ares system).

Large area lunar access—autonomous and/or human-assisted mobility. Several high-priority astrophysics programs are uniquely enabled by access to large areas of the lunar surface. Two such concepts are a large-area radio observatory located sufficiently far from human radio interference, and a widely dispersed retroreflector/transponder network to obtain increased accuracy for tests of general relativity. Both of these experiments/facilities could eventually require access to sites located hundreds to thousands of kilometers from a lunar base. Deployment of the assets potentially could be done either autonomously or via sortie missions by astronauts.

Evaluations and/or Trade Studies to Achieve Astrophysics Goals

Function of humans on the lunar surface. Although opportunities have been identified for astrophysics from lunar surface instruments that offer important science, these opportunities are either for small largely self-contained packages or for facilities (e.g., long-wavelength radio interferometers, lunar-ranging targets) that do not require precision alignment or positioning. Because general maintenance and servicing of such instrumentation may be uniquely enabled by hands-on access, a detailed assessment of the specific functionality of humans with respect to these opportunities should be done. This assessment can evaluate the viability of implementation plans that are entirely autonomous (or perhaps telerobotic), and to what extent such plans might compromise or enhance the performance of these facilities.

Options for large-area lunar surface emplacement. Two astrophysical observation opportunities require access to a large fraction of the lunar surface. First, a facility designed to observe the highly red-shifted hydrogen 21-cm line from the distant universe requires a significant amount of collecting area on the lunar far side. Current telescope designs envision a large number of individual elements (e.g., dipole antennas) that would need to be emplaced over this area. Second, sensitive tests of theories of gravity require laser retroreflectors, transponders, or both on the Moon. Optimal locations of these retroreflectors or transponders require wide spacing over the lunar surface at a variety of latitudes on the near side. An assessment is required of the manner in which these elements (dipole antennae or retroreflector/transponders) would be emplaced, and how their emplacement sites can be integrated with the long-term objectives of planetary and Earth sciences for global or complete nearside access.

Options for operations in free space. Capable operations in free space appear critical to achieving the major goals for science, industry, and national security. Assessments and trade studies are necessary to understand more fully how these operations may enable multiple national priorities and to provide a reliable basis for the design of elements of the lunar architecture. The assessment elements may include:
- the function in space of astronauts and robotic partners
- technology investment strategies
- options for coordinated development with industry, other Government agencies, and international partners
- design options for block changes to the Orion and Ares I/V systems
- cost estimates for possible augmentations to the exploration architecture
- cost trades related to lunar-based astrophysics observatories integrated into outpost activities vs. stand alone, independent space-based observatories
- traceability of in-space systems to major national goals

Strategies to maximize the potential for a meter-wave radio observatory. The expected signal strength from highly red-shifted hydrogen is quite small (~10 mK), requiring dynamic ranges of at least 1 part in 10,000, and the signal is expected to be spread over a significant frequency range, e.g., 10-200 MHz. To achieve such dynamic ranges and spectral access, a lunar telescope must be shielded from terrestrial, solar, lunar outpost, and other human-generated radio emissions. This requirement dictates a farside location;

however, even on the far side, multiple options exist to both realize the telescope and preserve the radio frequency environment. Examples of potential trade-offs include the degree of shielding and location on the far side, specifically with respect to how distant a long-wavelength observatory can be from a human outpost if relevant noise is being generated there; and the design of the communications infrastructure so as to maintain the radio frequency environment, particularly at low frequencies.

Capabilities of the Ares system. Future major missions in space, both for science and national security, can be enabled by the capabilities of the proposed Ares V heavy-launch vehicle, specifically the mass and volume the vehicle will be able to deliver to priority locations throughout the Earth-Moon system. Assessment and trade studies are needed to more fully understand how Ares V can enable multiple goals in space, and the science community must be informed about the performance capabilities of these vehicles.

Other findings. Astrophysics supports regular reviews (e.g., through the NASA Advisory Council structure) of major LAT decisions that may influence the science productivity of the lunar architecture. The VSE should be planned so as to include—and not preclude—capabilities that will enable astrophysics activities. Any lunar-enabled science can and should be evaluated and prioritized within the community by the decadal survey process. SMD funds are already committed to activities of the highest priority ranking in the decadal surveys. Assessments from the workshop do not replace or supersede the decadal survey process, although it is recognized that budgetary and operational considerations influence NASA's ability to implement specific objectives.

Earth Science

The goal of NASA Earth science research is to observe, understand, and model the Earth system in order to monitor its processes, discover the way changes occur, enable accurate prediction of these changes, and understand the consequences for life on Earth. The data currently used for this research is collected by an array of LEO and GEO (geostationary Earth orbit) satellite-based instruments. During the Workshop, there were two overarching questions addressed by the Earth Science Subcommittee (ESS) and interested members of the community:

1. What unique and/or complementary set of observations of the Earth can be made from the Moon that would significantly enhance data from LEO or GEO satellites?
2. Could those measurements be made from the notional lunar South Pole outpost on the rim of Shackleton Crater, and if not, from where could they be made?

A lunar Earth observatory would offer a unique, stable, serviceable platform for global, continuous, full-spectrum views of the Earth to address a range of Earth science issues over time, as well as provide instrument synergy among multiple LEO and GEO satellites for cooperative operations, enhanced calibration, and science. The proposed outpost location, while only offering limited views of the Earth, could still be useful initially for Earth science and instrument testing in the early stages of lunar exploration. However, a longer-term phased approach is desired, where the future observatory would be located away from the outpost in order to provide the desired continuous Earth views and collect time-dependent data of atmospheric composition, ecosystem health, and hazard monitoring. These data could be collected from a location further northward at a higher elevation near the outpost, from a location further southward, or in orbit at the Cislunar Lagrange Point (L1). The ESS also adopted the criterion of *unacceptable* if the Earth was in view less than 50 percent of the time; *acceptable* if the Earth was in view more than 50 percent of the time; and *desirable* if the Earth was in view more than 90 percent of the time.

The benefits of Earth observing from the lunar surface include a very stable platform that would be both accessible and serviceable, allowing a broader suite of instruments to be deployed and upgraded for Earth monitoring over a long time scale. The rotation of Earth as seen from the Moon would provide unprecedented and valuable temporal views of transient phenomena such as natural hazards, pollution, and climate. Furthermore, the Earth's orbital precession would allow observations of the polar regions (one pole at a time, in summer), which is not possible with GEO satellites. This dynamic observing opportunity was illustrated qualitatively by descriptions made by the lunar module pilot during Apollo 17's three-day trip to the Moon in 1972.

The potential for simultaneous measurements of the Sun and Earth from the Moon is another example of a set of observations that would allow a better understanding of the processes and interactions that determine the composition of the Earth's whole atmosphere, including the connections to solar activity. Such measurements would also map atmospheric species concentrations (greenhouse gases, aerosols, ozone) and provide real-time space weather data for predictive modeling of the space environment.

The concept of a lunar-based Earth observatory is highly compelling, but it must planned so as to maximize the science return while not distracting from critically needed Earth science observations from other platforms. Certain Earth Science observations can only be made well from LEO (e.g., high spatial resolution imaging, light detection and ranging [LIDAR], etc.), and these datasets should not be abandoned in the planning and implementation of a possible lunar-based Earth observatory.

To achieve the maximum return on Earth Science from the Moon, and to best integrate with the final lunar architecture, the ESS recommends a phased approach to instrumentation, beginning with relatively simple instruments deployed either at the surface by humans or into an L1 orbit, and extending to more complex instruments requiring significant infrastructure. A detailed assessment of this phasing concept is presented in the Earth Science workshop report in Appendix 2.

The following seven key points were made with respect to Earth Science and the exploration architecture:
1. **There are worthwhile and important Earth Science opportunities enabled from a lunar outpost.** Large telescopes are not needed; good science can be obtained with ~ 0.3 meter telescopes.
2. **Implementation of the VSE should be planned so as to accommodate capabilities that will enable Earth science.** Earth Science observations will become increasingly critical in the coming decades to record climate change, as well as to monitor and forecast natural hazards and potential disasters. Furthermore, there is great societal value to seeing Earth and its fragile atmosphere in the

vastness of space. However, without significant mobility, Earth observing science is limited at the notional Shackleton outpost location. Cost-effective alternatives to this location should be considered.

3. **Trade studies should consider options for outpost and observatory siting.** Engineering studies should be conducted to determine the best strategy for maximizing the Earth observation potential. The study of possible locations should include sortie locations within easy reach of the lunar outpost, such as a lower-elevation site with a clear view of Earth or a site at higher elevation and with Earth-facing slopes (e.g., Mt. Malapert) in regional proximity to the outpost. Both would possibly require new logistical and infrastructure considerations for the current lunar architecture.

4. **Studies are needed to model sensor designs and data quality needed to address the Earth science objectives from lunar platforms.** More precise and formalized engineering studies must be carried out in order to constrain both the common architecture desired in a future Earth observatory as well as specific sensor designs.

5. **Options for operations in free space should be developed.** Because of the limited options and cost associated with a lunar surface Earth observatory, an option would be to place instruments at the L1 point in order to provide full-Earth views and achieve the major Earth science goals. Assessments and trade studies are needed to understand how these operations might be enabled within the lunar architecture, as well as to understand the cost trades related to the options of either lunar-based observatories integrated into outpost activities or stand-alone, independent space-based observatories.

6. **Any science enabled by the lunar exploration should be evaluated and prioritized within the Earth Science community by the decadal survey process.** However, it is recognized that the recent Earth Science Decadal Survey was conducted without consideration of the future (lunar) exploration architecture. In the near term, continued activity within the NASA advisory structure will be required to fill this gap.

7. **Further involvement of the Earth Science community in planning for science enabled by the lunar exploration architecture is needed.** The ESS should organize and plan an "Earth Science from the Moon" workshop comparable to the recent "Astrophysics Enabled by a Return to the Moon" workshop.

Public Outreach. The human response to seeing Earth from space is significant. Images from the Apollo, Mariner 10, Voyager, Galileo, and MESSENGER missions have provided a global view of the home planet that can not be seen from LEO. It is important to expand beyond the occasional remarkable photograph of the entire Earth to a more systematic and synoptic set of measurements that can only be realized and enabled by the VSE. The lunar exploration architecture will miss a key opportunity in terms of public perception, outreach, and support if an outpost location is chosen with little or no opportunity to perform quantitative Earth observations and science.

Heliophysics

Presentations and discussions at the workshop made it apparent that the exploration architecture potentially available by NASA's return to the Moon presents interesting and exciting new opportunities to extend scientific progress in heliophysics in ways that have not been previously available or considered.

Since the inception of the space program with Explorer 1, and continuing through to the present space weather missions, scientists in the Heliophysics community have worked to develop a detailed understanding of the connected Sun-Earth-Moon system. The Moon is immersed in a plasma environment—the local cosmos—that is magnetized. These fields play an essential role in organizing the space environment and have significant influence on the terrestrial environment as well. It is the twisting and folding of the various interacting magnetic fields—of the Earth, Sun, and Moon—that regulate the local environment of the Moon and, therefore, the environment experienced by astronaut explorers. By working to understand this environment and, ultimately, predict the variations likely to occur from day to day and region to region, the efficiency, safety, and productivity of future lunar robotic and manned missions can be significantly enhanced.

The heliophysics science topics related to lunar exploration are grouped in four themes:
1. Heliophysics Science of the Moon—investigating fundamental space plasma processes using the Moon and its environment as a natural laboratory
2. Space Weather and Safeguarding the Journey—understanding the drivers and dominant mechanisms of the lunar radiation and plasma-dust environment that affect the health and productivity of human and robotic missions
3. The Moon as a Historical Record—seeking knowledge of the history and evolution of the Sun and solar system as captured in the lunar soil
4. The Moon as a Heliophysics Science Platform—exploring the possibilities of establishing remote sensing capability on the lunar surface to probe geospace, the Sun, and the heliosphere

Several issues that apply across Heliophysics science objectives are as follows:
- For several Heliophysics science opportunities, drop-off satellites or early robotic operations are optimal for deployment and thus, the availability and capabilities of an Apollo-like SIM bay is of great importance.
- Lunar science assessments formulated at this workshop are deemed to be valuable input to the next NRC Decadal Survey for Solar and Space Physics and NASA Heliophysics Science Roadmap. SMD has a well-validated process for establishing science priorities within their resource allocations. Once complete, the lunar science opportunities information should enter into this process in the same manner as other SMD pre-planning activities.
- Future evaluations of these science objectives must assess the cost effectiveness of these lunar site implementations versus more independent implementations that use robotic/unmanned missions around the Moon or elsewhere.
- For full mission success, many of these science objectives will necessarily require involvement of a scientist-astronaut as an integral part of the science experiment.

Heliophysics science objectives given a high priority include the following:
- Study the dynamics of the magnetotail as it crosses the Moon's orbit to learn about the development and transport of plasmoids
- Study the impact of the Moon on the surrounding plasma environment and incident solar wind to better understand the magnetotail

- Characterize the lunar atmosphere to understand its natural state. Of major importance are the electromagnetic and charged dust environment, and interaction with the variable space environment
- Map the surface electromagnetic field of the Moon to understand the operational environment of the Moon
- Characterize the dust environment at several locations on the lunar surface to better understand the operational environment of the Moon
- Monitor space weather in real time to determine and mitigate risks to lunar operations. Utilize the coordinated, distributed, and simultaneous measurements made by the heliospheric great observatory for predictive models of space radiation at the Moon
- Monitor lunar environmental variables in real time to determine and mitigate risks to lunar operations. Use real-time observations on the Moon to determine the potential and duration of radiation hazards, the electrodynamic plasma environment, and effects of dust dynamics and adhesion
- Understand the nature and history of solar emissions and galactic cosmic rays through studies of lunar regolith and regolith stratigraphy
- Perform meter-wave radio astronomy observations of the Sun to improve understanding of space weather
- Analyze the composition of solar wind to improve understanding of the composition and processes of the Sun. Composition and flux of interplanetary and interstellar grains should also be considered
- Image the interaction of the ionosphere and magnetosphere to understand space weather in the regions of space where most commercial and military space operations occur
- Perform high-energy and optical observations of the Sun to improve understanding of the physical processes of the Sun
- Analyze the Sun's role in climate change to gain a better overall understanding of climate

Not all of these objectives would necessarily be best achieved by an observatory at a polar outpost. For example, for real-time space-weather monitoring, upstream monitoring measurements must be located between the Sun and Earth and as close to the solar source as is feasible. Some of these objectives require multiple observation locations, and some require or are benefited by collocation with human operations. Some require a view of the Earth whereas others require maximum exposure to sunlight and solar wind. Detailed assessments are given for each of these objectives in Appendix 3.

Planetary Protection

The Planetary Protection Subcommittee (PPS) of the Science Committee of the NASA Advisory Council is charged with providing advice on planetary protection policy and mission categorization to NASA and the Planetary Protection Officer, in accordance with guidelines of Article IX of the 1967 Outer Space Treaty. At the Tempe workshop, the goal of the PPS was to ensure that planetary protection requirements for preventing biological and organic contamination of solar system bodies will be considered to the greatest extent possible during the development of technologies and procedures to enable human exploration of the solar system for which a return to the Moon is the first step.

By NASA policy, missions to the Moon are currently considered Category I, which means that operations on the Moon are not constrained by planetary protection restrictions on biological and organic contamination. The Moon is considered to be a sterile and organically clean environment, making it an optimal location to evaluate the magnitude and range of biological contamination associated with human exploration, and to develop technologies designed to mitigate contamination resulting from human presence. A better understanding of organic and biological contamination resulting from past or planned human activities on the Moon will facilitate the development and testing of equipment and technologies designed to limit human-associated contamination during exploration of more distant planetary bodies—including Mars—to which planetary protection restrictions are applied.

Planetary Protection, Key Findings

Below are the three key findings regarding planetary protection that emerged from the workshop. Addressing issues one and two during exploration of the Moon were considered essential to preparations for future missions to Mars. The third concern, specific to the Moon, is that exploration of scientifically interesting polar regions on the Moon does increase the possibility of contamination that might interfere with future scientific discovery. These key findings are as follows:

1. Exploration of the Moon has produced and will produce biological and organic contamination at the sites where human and/or robotic exploration takes place. Operations on the Moon are not constrained by planetary protection restrictions, which makes the Moon an optimal location to establish the magnitude of contamination associated with human exploration. Previous lunar exploration efforts, including both robotic missions and the manned missions of the Apollo program, have left behind artifacts on the Moon that contain organic and microbial contaminants. These locations are ideal for testing planetary protection technologies and procedures to detect biological or organic contamination. In addition, the Moon is an excellent testbed for developing and testing technologies for containment of collected samples in order to prevent both forward contamination of the sampling site and backward contamination of the habitat, the return vehicle, and the laboratory in which the sample containers are to be opened.

2. Many processes and technologies required for planetary exploration are likely to produce organic and biological contaminants that are regulated by planetary protection policy. Because organic and biological contamination of the Moon is not restricted, technologies that will be required for exploration of protected locations can be tested and optimized without costly limitations. The Moon is expected to be an excellent testbed for developing the technologies required to permit human exploration of protected planetary bodies. The necessary technologies that will need optimization to minimize contamination include pressurized habitats and spacesuits, as well as robotic and human-associated mobile equipment used for exploration or in situ resource utilization. Such technologies and procedures are absolutely required before humans can be permitted to travel to Mars or other protected solar system bodies.

3. Lunar volatiles in polar deposits are susceptible to organic contamination during exploration, and future investigation may indicate that these regions contain materials of interest for scientific research. These regions of the Moon, though currently considered Category I, may become protected at a greater level pending future policy discussions.

Planetary Protection Objective Assessments

The two main science objectives considered by the PPS were the evaluation of astrobiology protocols and measurement technologies that will be used to test for life on other planets, and the development of planetary protection protocols as part of the next generation planetary protection policy. Both of these objectives can be accomplished at an outpost location and within the notional lunar exploration architecture. These two objectives were subdivided to highlight or to expand specific components or activities, and these were each assigned priorities as follows:

High Priority
- In order to assess the contamination of the Moon by lunar spacecraft and astronauts, perform in situ investigations of a variety of locations on the Moon using highly sensitive instruments designed to search for biologically derived organic compounds
- Understand possible contamination of lunar ices with non-organically clean spacecraft. Evaluate and develop technologies to reduce possible contamination of lunar ices
- Prior to planning human Mars missions, use the Moon, as well as lunar transit and orbits, as a testbed for planetary protection procedures and technologies involved with implementing human Mars mission requirements

Medium Priority
- Perform chemical and microbiological studies on the effects of terrestrial contamination and microbial survival during both lunar robotic and human missions (dedicated experiments and "natural" experiments in a variety of lunar environments/depths, etc.), including effects generated during the Apollo missions (study Apollo sites)
- Develop technologies for effective containment of samples collected by humans that will help prevent forward and backward contamination during Mars missions

Low Priority
- Use the lunar surface as a Mars analog site to test proposed life detection systems in a sterile environment for future use on Mars

Enabling Technologies
Technology development is needed to ensure that life support and habitat technologies to be used for later human missions to other solar system bodies that have more stringent planetary protection requirements. Technologies and instruments developed for robotic spacecraft exploration and adapted for human interface—either with the assistance of a robot or through direct human operation while wearing a spacesuit—include the tools for sample collection and sensitive, rapid assay methods using field-portable equipment. These should be reinvestigated for relevance to human exploration requirements. Commercial off-the-shelf technologies, however, are not rated for spaceflight, and necessary modifications may require re-engineering to accommodate human-rated spaceflight requirements such as low outgassing from construction materials and radiation-resistant electronics. The Moon can well be used as

a testbed of advanced life-support systems for Mars exploration, emphasizing sustainable high efficiency closed-loop systems and comprehensive efforts to assess their effectiveness.

Planetary Protection Issues

The near-term focus on exploration of the Moon affords a unique opportunity for testing planetary protection protocols in a challenging space environment that is known to be sterile but is not restricted by planetary protection policy. Every effort should be made to take advantage of this opportunity in order to ensure that planetary protection protocols are established to the extent that will be required for future human missions to solar system bodies receiving more than Category I protection.

A separate, follow-on meeting to explore opportunities in biological sciences in partial gravity and at a pressurized lunar outpost is suggested. Such a meeting will continue and expand the recent effort that brought together planetary protection experts, astrobiologists, life-support specialists, and engineers to discuss human exploration of space.

Substantial proportions of lunar dust are submicron-sized and could pose a significant health hazard. Current efforts to use data from Apollo and terrestrial dust exposure studies should be strongly encouraged to better understand exposure times, particle distributions, particle morphology, chemistry, and reactivity that may pose health risks.

Planetary protection technologies to reduce contamination from human missions must be supported if human missions to Mars are to be planned and implemented with appropriate planetary protection protocols.

Effective communication with the public about planetary protection goals and requirements is key to garnering and retaining public support for both robotic and human missions to other planetary bodies.

Planetary Science

The Planetary Sciences Subcommittee grouped science objectives under five broad science themes as follows:
1. Investigation of the geological evolution of the Moon and other terrestrial bodies, including the origin of the Earth-Moon system
2. Improved knowledge of impact processes and impact history of the inner solar system
3. Characterization of regolith and mechanisms of regolith formation and evolution
4. Study of endogenous and exogenous volatiles on the Moon and other planetary bodies
5. Development and implementation of sample documentation and return technologies and protocols

Within the context of these five science themes, 16 specific science objectives emerged as follows:*

1. **Determine the internal structure and dynamics of the Moon to constrain the origin, composition, and structure of the Moon and other planetary bodies.** (mGEO-1) Achieving this objective requires emplacement of a seismic network with long-lived power supply for seismometers and three or four widely separated sites because this objective cannot be addressed entirely from a single site. However, a seismic station (geophysical station) should be set up at an outpost site because it would provide some information about the interior and, importantly, it would represent a start toward establishing a long-duration global seismic/geophysical network. This objective is one that would benefit from collaboration with international partners if they have landed missions to other lunar locations and, therefore, could emplace additional network nodes.

2. **Characterize impact flux over the Moon's geologic history to understand early solar system history.** (mGEO-7) This objective requires the return of geologic samples for precise age dating by isotopic methods. Long-range surface mobility and/or access to multiple crater locations (e.g., via sorties) are needed to obtain the range of samples required to adequately determine the impact flux. If the outpost was located within a large basin not previously sampled, significant progress could be made. For example, if the site were inside the South Pole-Aitken basin, it would be possible to sample the basin melt sheet (hence, to be able to date the event) and to determine the ages of superposed younger basins. Access to South Pole-Aitken basin requires a far-side, southern hemisphere site.

3. **Determine the composition and evolution of the lunar crust and mantle to constrain the origin and evolution of the Moon and other planetary bodies.** (mGEO-2) Achieving this objective requires targeted sample returns from multiple locations; however, some progress can be made by intensive study of one site as well as by documentation and return of rock and regolith samples collected throughout the region surrounding the outpost. How much progress can be made depends on the geological setting of the specific site chosen; proximity to a diversity of geologic terrains is particularly important.

4. **Study the lunar regolith to understand the nature and history of solar emissions, galactic cosmic rays, and the local interstellar medium.** (mGEO-9) Activities needed to accomplish this objective include drilling and/or trenching of the lunar regolith. Extensive regolith excavation at a single site could address this objective by identifying layers deposited by specific impact events; however, such activities would be best done where interlayered volcanic deposits provide an age record. Extensive ISRU processing could aid this scientific activity.

5. **Characterize the lunar geophysical state variables to constrain the origin, composition, and structure of the Moon and other planetary bodies.** (mGEO-3) These variables include the gravitational potential field, heat flow, lunar rotational fluctuations, lunar tides and deformation, and the present and historic magnetic fields. Little progress can be made on this objective from a single site, with the exception of temporal heat flow and magnetic measurements, which should span the lifetime of the outpost. The utility of a single heat-flow measurement depends on the complexity of the geological setting of the site.

---

* This is an approximate order of priority. See Appendix 5 for specific priority rankings.

6. **Characterize the crustal geology of the Moon via the regolith to identify the range of geological materials present.** (mGEO-5) This approach is less effective than going to diverse terrains on the Moon to sample the crust, but significant progress can be made at one site. A polar location represents a previously unsampled terrain. Regolith samples and rock fragments in the regolith complement any collection of large-rock samples. Regolith sampling could be conducted robotically.

7. **Characterize the impact process, especially for large basins on the Moon and other planetary bodies, to understand this complex process.** (mGEO-6) Significant progress can be made at a single site by studying a number of impact craters in detail; however, local to regional surface mobility for astronauts is needed. Achieving this objective requires orbital and sample data, including geological and geophysical field studies and return of key samples to Earth.

8. **Characterize lunar volatiles and their source to determine their origin and reveal the nature of impactors on the Moon.** (mGEO-12) The analysis of volatiles in the lunar exosphere and in/near polar cold traps are well-enabled by a polar outpost location. In terms of phasing, this activity should be done early in the human exploration program.

9. **Determine the origin and distribution of endogenous lunar volatiles as one input to understanding the origin, composition, and structure of the Moon and other planetary bodies.** (mGEO-4) Achieving this objective requires landing sites with the best chance of yielding significant information about lunar endogenous volatiles, such as pyroclastic deposits, near volcanic vents, or sources of possible recent outgassing.

10. **Investigate meteorite impacts on the Moon to understand early Earth history and origin of life.** (mGEO-8) This objective is aimed at finding Earth or other extralunar materials ejected from large impacts on Earth or by collisions involving other objects that later fell to the Moon. This objective requires access to multiple impact craters and regolith samples. It is well addressed at a single outpost site where large amounts of regolith can be processed and techniques employed to search for key indicator minerals or chemical compositions that would indicate the origin of the impactor.

11. **Determine lunar regolith properties to understand the surface geology and environment of the Moon and other airless bodies.** (mGEO-10) Achieving this objective involves extensive study of regolith, including excavation, sampling, and geophysical studies. This objective can be achieved at an outpost site. The investigation would go far beyond what is known from Apollo cores and active seismic measurements, and could involve in situ measurements of many geotechnical and other regolith properties. Such investigations would be enabling for exploration.

12. **Characterize the lunar regolith to understand the space weathering process in different crustal environments.** (mGEO-11) This requires local surface mobility, trenching, sample documentation, collection, and return of samples to Earth. It can be done well at a single site with detailed investigation of regolith at different proximal locations and with different degrees of surface exposure.

13. **Characterize transport of lunar volatiles to understand the processes of polar volatile deposit origin and evolution.** (mGEO-13) This objective is best approached through global access (range of latitudes and locations). Much of this objective, however, can be achieved at a polar outpost site through access to permanently shaded craters and regolith near to, and at a range of distances from, the pole.

14. **Characterize volatiles and other materials to understand their potential for lunar resource utilization.** (mGEO-14) Ground truth/in situ characterization of deposits located from orbital data can lead to accurately targeted locations on the Moon. This should be done during the robotic precursor phase to identify the best outpost location since conducting this activity from a polar outpost location instead of during the precursor phase will adequately characterize the deposits at the site, but would be too late to influence the optimal outpost location. Therefore, this should be considered an exploration-enabling objective/activity.
15. **Provide curatorial facilities and technologies to ensure contamination control for lunar samples.**** (mGEO-15) This objective can be well achieved at an outpost location. Potential polar volatile deposits would provide a test case for extremely environmentally sensitive sample documentation, collection, transfer, and processing.
16. **Provide sample analysis instruments and protocols on the Moon to analyze lunar samples before returning them to Earth.**** (mGEO-16) This objective can be achieved at an outpost and could prove useful to enable adequate sample return in the event of return-mass limitations. Instrumentation can be used by astronauts to aid in documentation and selection of geologic samples.

Planetary Science Recommendations

Geophysical Networks. Achievement of several of the highest-ranked lunar science objectives requires the deployment of long-lived geophysical monitoring networks. Precursory technology investments are needed, such as the development of a long-lived power source and a deployment strategy for stations that are part of such networks. Networks could be built up in partnership with other space agencies provided that a framework for compatible timing and data standards is established. The tradeoff between station lifetime and the timeframe for network deployment should be fully explored.

Sample Return. Achievement of several of the highest-ranked scientific objectives requires the development of a strategy to maximize the mass and diversity of returned lunar samples. The Planetary Science Subcommittee (PSS) views the 100 kg total return payload mass allocation in the current exploration architecture for geological sample return as far too low to support the top science objectives. The PSS requests that Curation and Analysis Planing Team for Extraterrestrial Materials (CAPTEM) be asked to undertake a study of this issue with specific recommendations for sample return specifications to be completed as soon as possible. The PSS recommends that NASA establish a well-defined protocol for the collection, documentation, return, and curation of lunar samples of various types and purpose in order to maximize scientific return while protecting the integrity of the lunar samples.

Astronaut Training. As part of the developing lunar exploration architecture, extensive geological, geochemical, and geophysical field training should be established as an essential component in the preparation of astronaut crews and the associated support community for future missions to the Moon. Training should involve experts and experience from the non-NASA community, as well as NASA personnel of

---

** These objectives relate to implementation activities and are ranked, along with the other science objectives, with high priority.

significant background and experience in field exploration and space mission planning and execution. The training program developed for the Apollo 13–17 missions should be considered a starting point for training of the next generation of lunar explorers. Crews for future lunar missions should include astronauts with professional field exploration experience. Research is needed to determine the best use of robots to assist humans in activities associated with the lunar architecture.

Mobility. To maximize scientific return within the current exploration architecture, options should be defined and developed for local (~50 kilometers), regional (up to 500 kilometers), and global access from an outpost location. It is important that access to scientifically high-priority sites not be compromised by mobility limitations, both for outpost and sortie missions.

Robotic Missions. Robotic missions are highly desirable to carry out many of the highest-priority lunar-science objectives. Robotic precursor missions beyond LRO are important for both basic and exploration science (e.g., determining seismicity in proposed outpost locations and defining the nature of the cold-trap volatile deposits). To achieve the highest-ranked lunar science objectives, continued robotic sortie missions will be needed both before and after human presence is established.

CEV-SIM Bay. The CEV should have a capability similar to the Apollo SIM to facilitate scientific measurements and the deployment of payloads from lunar orbit.

Landing Site and Other Operational Decisions. Scientific input should be an integral component of the decision-making process for landing-site targets and for exploration planning and execution for a lunar outpost or any lunar mission.

Integration of Data Sets. Lunar data sets from all past missions, LRO, and future international missions should be geodetically controlled and accurately registered to a common format that will facilitate the creation of cartographic products that, in turn, will enable landing-site characterization, descent and landed operations, and resource identification and utilization through a variety of data-fusion techniques.

Technology Developments. A lunar instrument and technology development program is needed to achieve several of the highest-ranked scientific objectives such as exploration and sample documentation aids, long-lived 1-10 W power supplies, deployment of networks from orbit (e.g., from the CEV-SIM bay), sampling in permanently shadowed regions, and development of robotically deployable heat-flow probes.

Sustained Scientific Input to Lunar Exploration Planning. Regular reviews of the major decisions that will influence the science outcome and legacy of lunar exploration should be carried out by the Council and its science subcommittees, with their findings and recommendations transmitted to NASA. Topics for such reviews should include:

- Options for full access to the Moon (low, mid, and high latitudes; near side and far side; polar)
- Pre- and post-landing robotic exploration opportunities and missions
- Options to mix human and robotic exploration
- Surface science experiments and operations at the human outpost
- Surface science experiments and operations during human sorties
- Mission planning
- Critical items in space hardware design, including:
    - delivery of science experiments to the lunar surface
    - returned payload constraints and upload of science (samples, data) from the lunar surface
    - orbiting module science requirements (e.g., SIM bay)
    - crew orbiting science operational requirements (e.g., portholes)
    - mission control science requirements during operations

FINDINGS OF THE WORKSHOP

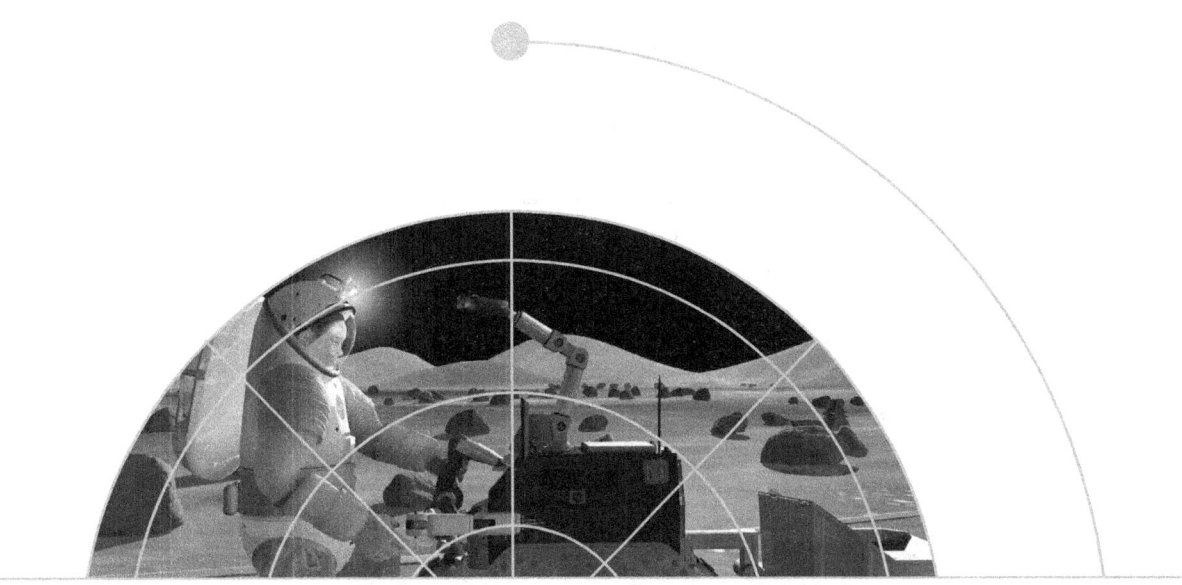

OUTREACH MESSAGE AND HIGHLIGHTS
OF THE WORKSHOP

During the workshop, the Outreach Committee formulated messages relative to each of the subcommittee disciplines, both for the science community and the public. These messages provide an excellent summary of the scientific possibilities associated with or enabled by the return to the Moon and are given in the following paragraphs.

Astrophysics

Key messages for the science community
1. The return to the Moon will enable progress in astrophysics through the associated infrastructure. Some important astrophysical observations, as well as a few smaller experiments, can be uniquely carried out from the lunar surface and in lunar orbit. Potentially important observations include long-wavelength radio observations from the far side of the Moon, lunar laser ranging observations for fundamental physics, and characterization of Earth and dust in the solar system as they apply to extra-solar planet research.
2. Astronauts can carry relatively small astronomy experiments with them to the Moon. These packages can accomplish a wide range of science, from determining how gravity really works to understanding how to search for signs of life on other worlds by using the full view of our own Earth.
3. The rockets that will take us back to the Moon give astronomers the heavy lifting they need to put bigger and better telescopes into space. Among other things, these telescopes will look for Earth-like planets beyond our solar system, investigate the environment around black holes, and probe the dark energy that makes up most of our universe.

Key messages for the public
1. The far side of the Moon provides a radio-quiet zone that enables astronomers to look back in time and find out when the first stars were born.
2. Astronauts can carry relatively small astronomy experiments with them to the Moon. These packages can accomplish a wide range of science from understanding how gravity really works to using the full view of our own Earth in understanding how to search for signs of life on other worlds.
3. The rockets that will take us back to the Moon give astronomers the heavy lifting they need to put bigger and better telescopes in space. Among other things, these telescopes will look for Earth-like planets beyond our solar system, investigate the environment around black holes, and probe the dark energy that makes up most of our universe.

Earth Science

Key messages for the science community
1. A lunar observatory provides a unique, stable, and serviceable platform for continuous global full-spectrum observation of the Earth, which will allow researchers to address a range of Earth science issues over the long-term.
2. Synergy of current LEO, GEO, and Global Positioning System (GPS) assets with lunar instrumentation will insure the collection of the widest array of information from a lunar base.
3. There are numerous atmospheric profiling opportunities from visible (stars) to microwave (GPS) to VHF (communications).

Key messages for the public
1. The view from the Moon offers a unique perspective of the full Earth all at once and over time.
2. From an Earth observatory on the Moon, we can take the pulse of the Earth by monitoring long-term Earth events such as climate variability, air pollution sources and transport, natural hazards (extreme weather, volcanic plumes, hurricanes), and seasonal and long term variations in polar ice.
3. By viewing the Earth from a distance, we can collect data to help us detect and study distant Earth-like planets.

Heliophysics

Key messages for the science community
1. Understanding our space environment is the first step to safeguarding the journey.
2. The Moon can be used as an unique vantage point to better understand the Sun-Earth space environment—our home in space.
3. The analysis of lunar regolith will provide a history of the Sun's brightness and radiation output, in addition to revealing how the Sun-Earth connection has changed over time.
4. The Moon is a natural laboratory for space physics.

Key messages for the public
1. The same key messages above apply to the public as well as scientists.
2. In terms of safeguarding the journey, we must recognize that outer space is a perilous ocean through which we must travel to reach the dusty shores of the Moon, then Mars beyond that. Space is permeated with charged particles, electromagnetic fields, and blasts of radiation from the Sun; therefore, we seek to enhance astronaut and robot productivity and safety by forecasting space weather and charged-particle impacts while also learning to mitigate their results.

Planetary Protection

Planetary Protection is an important ongoing focus of both science research and mission planning to safeguard planetary environments and exploration throughout the solar system.

Key messages for the science community
1. Based on the Outer Space Treaty, international policies, and decades of research and experience on protecting planetary bodies during exploration, lunar missions will not require special planetary protection controls.
2. Lunar exploration provides the opportunity for an integrated test bed of sophisticated technologies and methods needed to understand and control mission-associated contamination on long-duration expeditions.
3. Lessons learned on the Moon will provide essential, enabling, and comparative information, such as understanding background and mission-associated organic and inorganic contaminants to ensure protection of planetary environments and humans as we explore Mars and other destinations.

Key messages for the public
1. Based on international treaties, policies, and decades of research experience on protecting planetary bodies during exploration, lunar missions will not require special planetary protection controls.
2. Lunar exploration provides a good opportunity for testing technologies and methods to understand and control mission-associated contamination on long-duration expeditions.
3. Lessons learned on the Moon will provide essential information to ensure protection of planetary environments and humans as we explore Mars and other destinations.

Planetary Science

Key messages for the science community
1. The Moon is critical for accessing the early formation, differentiation, and impact history of the terrestrial planets, with implications for biotic evolution of Earth and, potentially, Mars.
2. Additional data are needed: geophysical and geochemical data to determine the composition, structure, condition, and evolution of the lunar interior; data from the lunar surface to understand the processes that have occurred during its evolution, such as the history of impact cratering and formation of regolith, and the distribution of resources; and data to inform us more about the lunar environment (conditions in cold traps, atmosphere, volatiles).
3. New data will enable us to validate lunar science process models, understand the early history and evolution of Earth and other terrestrial planets, and prepare for human habitation of the Moon and beyond. Furthermore, the notional exploration architecture as presented (access to South Pole-Aitken Basin from the southern rim) will enable long-term lunar science in a region of high interest, and can potentially address several scientific questions (e.g., crust to upper-mantle access, impact processes). The scientific goals will have to be prioritized in a cohesive vision across a timeline. This long-term planning should encompass (1) robotic and robotic/human sorties to acquire distributed samples and establish the geophysical network necessary to prepare for a lunar outpost, as well as to address the fundamental science questions; and (2) samples from diverse locations on the lunar surface and subsurface to address fundamental science questions. In-situ science will optimize science output/return. The exploration and science community should actively participate in the development of human capital to fuel the pipeline of scientists and engineers.

Key messages for the public
1. The Moon holds a record of the early history of terrestrial planetary formation and change that is absent on other planets because they have undergone active resurfacing processes such as weathering and plate tectonics.
2. We are in a position to build on four decades of lunar science. There is much more new information to learn about our Moon and—from the Moon—about the Earth. For example, the Moon maintains a cratering history that may inform our understanding of the evolution of life on Earth and potentially elsewhere in the solar system.
3. The lunar outpost will serve as a testbed for science and exploration of the Moon, Mars, and beyond (camp first in your own back yard!).

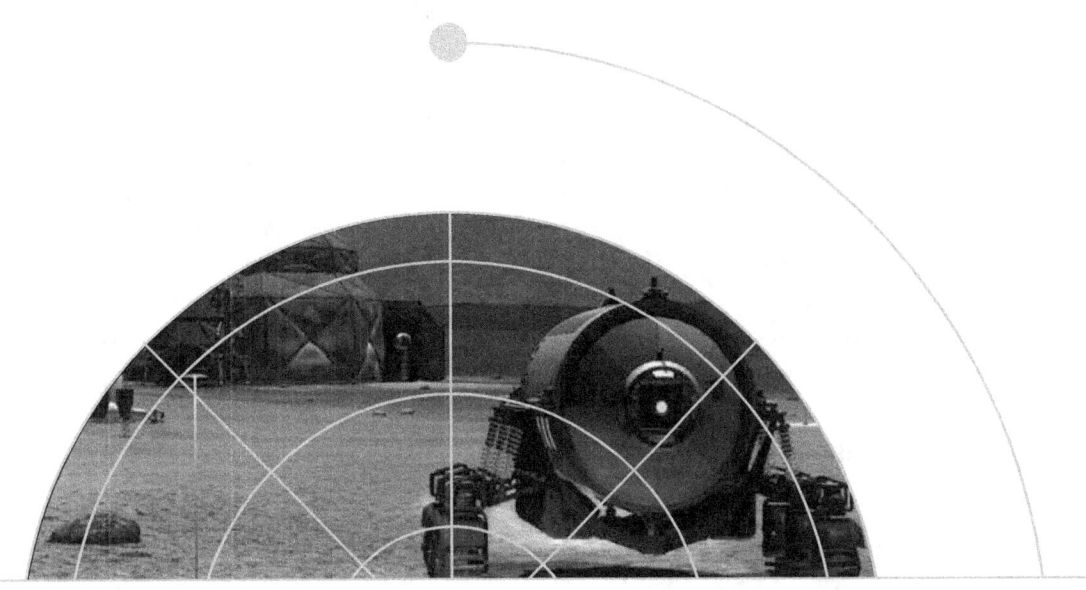

CONCLUDING STATEMENT

An outpost on the Moon will help us understand our "home in space" while providing a beginning to the next steps toward sustained human presence on other planets.

An outpost on the Moon will enable many scientific observations and activities that will address fundamental questions in space science. Through scientific components of our exploration, we seek to understand how and why the Sun varies and what effects these variations have on the Earth, not just for the present, but over long periods of time as well. How do the Earth and other planets respond to changes in the Sun's activity and to other solar system events such as the impact of asteroids and comets? What is it about the Earth-Moon system that makes our part of the solar system, and Earth in particular, perhaps uniquely habitable? What changes have occurred over time on the Moon, Earth, and other planets that affect the ability of life to claim a foothold and then sustain its presence? How unique is our solar system within the universe, and how did our solar system and galaxy come to be as they are? Armed with a better understanding of our planetary past and our place in the universe, humanity will be richer in knowledge and better able to chart a course into the future. Scientific roles within the exploration architecture are key to charting this course and, therefore, to implementation of NASA's Vision for Space Exploration.

List of References

National Research Council Interim Report on The Scientific Context for Exploration of the Moon—Interim Report, The National Academies Press, 2006.

Exploration Systems Architecture Study—Final Report, NASA TM 2005 214062, November 2005; *http://www.nasa.gov/mission_pages/exploration/news/ESAS_report.html*.

Lunar Exploration Analysis Group Science Activities and Site Selection Specific Action Team Rapid Response Report, July 11, 2005; *http://www.lpi.usra.edu/leag/reports/SASS_SAT_Report_7-12-05.pdf*.

Field Exploration and Analysis Team—Planetary Field Exploration Project White Paper: Revision Marh 30, 2007.

White Paper and Presentation Archive: An electronic archive of white papers (oral, poster, and print only) and workshop presentations is maintained by the Lunar and Planetary Institute at *http://www.lpi.usra.edu/meetings/LEA/*.

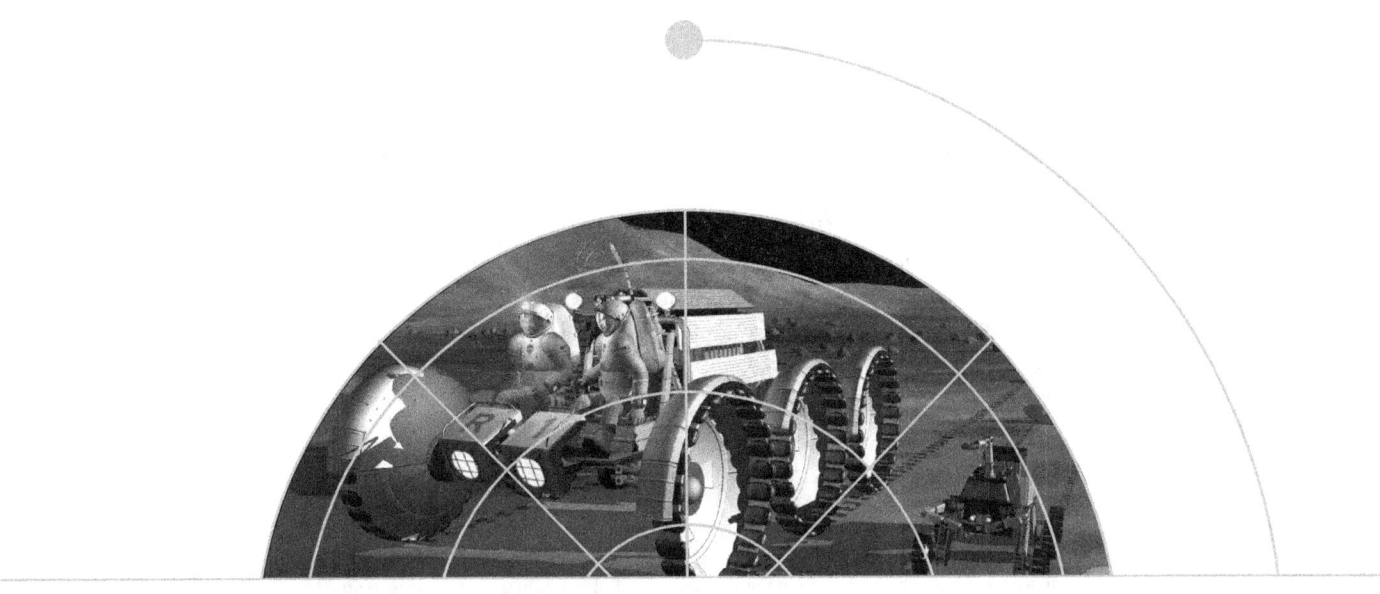

APPENDIX 1: WORKSHOP FINDINGS
ASTROPHYSICS SUBCOMMITTEE

In November 2006, representatives from the U.S. astrophysics community participated in a workshop entitled "Astrophysics Enabled by the Return to the Moon." The workshop was organized by the Space Telescope Science Institute (STScI), in collaboration with the Johns Hopkins University, the Association of Universities for Research in Astronomy, and NASA. The decision to hold the meeting was in direct response to the encouragement by the NASA Administrator to provide scientific input to the VSE, which envisions the return of humans to the lunar surface by 2020. The STScI workshop focused primarily on science. The broad workshop goal was to identify key questions in astrophysics, and to critically examine whether the proposed return to the Moon can—either directly or through the capabilities developed by the VSE—provide opportunities for significant progress toward answering those questions. Four science goals were identified that are widely believed to pose intriguing astrophysical challenges for the next two decades, and to encompass the breadth of current astrophysics research. These are (in no particular order):

1. What is the nature of the dark energy that is propelling the cosmic expansion to accelerate?
2. Are there habitable extrasolar planets and, in particular, is there extraterrestrial life?
3. Which astronomical objects and which physical processes were involved in the "first light" in (and the re-ionization of) the universe?
4. How did galaxies and the large-scale structure of the "cosmic web" form?

The participants in this Tempe workshop agreed with these scientific goals, and adopted them as a framework within which to evaluate the objectives crafted by the Lunar Architecture Team. The Tempe workshop had two tasks: (1) confirm that the list of objectives identified by the Lunar Architecture Team was complete and representative of the science goals outlined above; and (2) through invited presentations, posters, and general discussion, assess the capabilities of the lunar architecture to achieve those objectives. The assessments include both intrinsic scientific value and also the workshop attendee's best understanding of how well the objectives meshed with the architecture as we understood it. Attendees were also asked to identify key technology developments required for implementation, as well as needed trade studies.

Here we summarize our results. First, we present our key findings regarding astrophysics as enabled by the lunar architecture. We then list enabling technologies, along with succinct discussions of why those technologies were identified; we also identify a "Point of Contact" for each technology. After that, we list the studies relevant to the highest priority objectives. Finally, we provide a table that identifies each LAT objective, provides our assessment, discusses the primary factors that motivated the assessment, and details the specific trade studies associated with each objective.

Key Findings for Astrophysics

1. **There are some worthwhile astrophysical opportunities within the lunar architecture.**
   The most promising opportunities seem to be low-frequency radio telescopes on the lunar surface, which have a reasonable science and technology expansibility from small precursors to eventually large facilities. Also, in that field there are good synergies between heliospheric physics and

astrophysics. Smaller "payloads of opportunity" can also provide interesting and competitive science without deleterious effects on SMD planning or budget. These smaller payloads, which should be competitively selected, do not necessarily do science of the highest decadal survey priority, but they still do good science that meshes well with the lunar architecture. We recommend regular reviews through the NASA Advisory Council of major LAT decisions that may influence the science productivity of the lunar architecture.

2. **VSE should be planned so as not to preclude—and, to the extent possible, include—capabilities that will enable astrophysics.** This finding refers both to possible additions of capability in the future and to keeping environments in an appropriate condition for future development.

3. **Any lunar-enabled science can and should be evaluated and prioritized within the community by the decadal survey process.** SMD funds are already committed to activities of the highest priority ranking in the decadal surveys. Our assessments should not be considered to in any way replace or supersede the decadal survey process. The assessments include, in addition to intrinsic science, the manner in which the science may mesh with the lunar architecture.

Detailed Assessment of LAT Astrophysics Objectives

Key for Assessments (details provided in the "Comments" column for each objective):

1 = High priority science and/or a perceived excellent mesh with lunar architecture
2 = Medium priority science and/or difficult fit with lunar architecture
3 = Low priority science and/or poor fit with lunar architecture

Key for Trade Studies (details provided in the next section):

[1] Function of humans on lunar surface
[2] Options for large-area lunar-surface emplacement
[3] Options for operations in free space
[4] Strategies to maximize the potential for a low-frequency observatory
[5] Capabilities of the Ares system

Code	Title	Assess	Comments	Studies
mA1	Low-Frequency Radio Observations	1	A low-frequency observatory on the lunar farside would open a new window below the ionospheric cutoff. Such a facility would have exciting applications in cosmology, extra-solar planet characterization, and the physics in the nuclei of active galaxies. There are good opportunities for scientific and technological expansibility, as well as strong synergies with some heliospheric experiments.	[1],[2],[4]
mA2	Lunar Optical Interferometer	3	Space-based telescopes will do a better job of covering the UV plane. Free space is also a cleaner and more flexible environment	
mA3	Detect Gravitational Waves	3	Free space is a superior environment.	
mA4	Large Lunar Optical Telescope	3	Transit telescopes have limited scientific usefulness. Free space is a cleaner and more flexible environment.	

mA5	Lunar Energetic Observatory	1 3	1 = Low-Earth orbit mission: the Ares V would uniquely enable this. Potential successor mission to the Gamma Ray Large Area Space Telescope (GLAST). 3 = On the lunar surface: this option would require significant in situ construction capabilities (125 tons of materials processed on surface). Alternative of using Ares V to launch detector to low-Earth orbit seems more attractive.	[5]; [1] for lunar surface option
mA6	Search for Exotic Stable States of Matter	3	There are already very strong limits from terrestrial studies. The science case was not compelling.	[C]
mA7	Fundamental Physics	1	A multispectral sensor ranging from the UV to the TIR (much like the Moderate Resolution Imaging Spectroradiometer [MODIS]) could fulfill several objectives on this list (mEO3, mEO5 and possibly mEO4, mEO12). Full Earth views are critical, but the telescope could be as small as 30-50 cm (hence, the improved [2] / [1] ranking).	[1]
mA8	Near-Earth Asteroids	N/A	Sent to Planetary Science Subcommittee.	[A]
mA9	Site Characterization	1	The highest priority for site characterization would be for a low-frequency radio observatory. Many astronomical applications need a clean environment, and there is also good synergy with site characterization activities in the other disciplines (heliophysics, planetary protection).	[1],[2],[4]
mA10	"Piggyback" Missions to Surface and Lunar Orbit	1	A good fit with lunar architecture. This capability offers the potential for frequent, inexpensive access to space. A science assessment would depend on the specific competitively selected mission. There is a wide range of potential applications, including simple retroreflectors, Earth-observing telescopes, and inner zodiacal dust characterization (the latter two concepts have implications for extra-solar planet research).	[1],[5]
mA11	Large Telescope at Earth-Sun L2	1	Ares V provides a launch vehicle capable of launching an 8- to 15-m optical/UV space telescope. This capability would remove pressure for light-weighting of structures and optics. Other possible payloads include infrared, x-ray, and gamma-ray telescopes. Fairing sizes of 12 meters have been identified as useful.	[3],[5]

Enabling Capabilities for Astrophysics

Examples of capabilities that will enable astrophysics. Within each category, no prioritization is implied.

High priority for astrophysics, may influence architecture

Radio-quiet (RFI) environment and infrastructure on the lunar far side, or near the Shackleton site, for a low-frequency observatory (e.g., the local lunar atmosphere and electronic density goes up significantly for a month with every landing). Point of contact: Joe Lazio (NRL)

The farside of the Moon, because of its shielding from terrestrial and solar radio emissions and its lack of a permanent ionosphere, offers the potential for extremely sensitive probes of the cosmic evolution of the universe. In the "hot Big Bang" cosmology, the universe started in a dense, ionized state. As it expanded and cooled, it underwent a transition to a neutral state in a process is called recombination. After recombination, baryons began to collapse into regions of higher density, leading to the formation of stars and galaxies. Today, their radiation maintains the universe in an ionized state. During at least a portion of this process of structure formation, the dominant baryonic component of the universe, hydrogen, should have emitted 21-cm radiation. If this radiation can be detected, the (highly) red-shifted 21-cm signal will provide a unique and sensitive probe of cosmic evolution, including the formation of the first structure in the universe and the first luminous objects. The implied wavelength range (wavelength > 1.5 meter or frequency < 200 MHz) is a heavily-used spectral region on Earth (e.g., for FM radio). The expected strengths of the hydrogen 21-cm signals are quite small—many orders of magnitude below the strength of typical human-generated transmissions, solar radio emissions, and natural terrestrial radio emissions. Thus, the most sensitive observations of these red-shifted 21-cm hydrogen signals will be obtained in a location that is shielded from such interfering signals. The lunar far side is an excellent environment for these studies.

Large launch vehicles capabilities—VSE will include large launch vehicles such as the Ares V, and the science community should be part of the dialogue in crafting the launch vehicle capabilities or those derived from it (examples include but are not limited to volume, large mass capability, and similar aspect ratio). The community can envision several large telescopes which could utilize this capability. Point of contact: Phil Stahl (NASA/MSFC)

The Ares launch system (i.e., Ares I and Ares V) offers a capability that could revolutionize astrophysics (and other sciences) by enabling entirely new classes of missions that will achieve priority astrophysics. Specifically, current estimates for the launch mass and faring volume could enable: (1) a 6- to 8-meter class monolithic UV/Visible/IR observatory, (2) a 5-meter cube (130,000 kilogram) gamma ray water calorimeter, (3) a 4-meter-class x-ray observatory, (4) a 15- to 20-meter-class far-IR/sub-mm observatory, (5) a 25- to 30-meter-class segmented UV/Visible/IR observatory, (6) a 150-meter-class radio/microwave/terahertz antenna, or (7) constellations of formation-flying spacecraft.

Capability for secondary payload of small or medium science instruments (on lunar orbiters, or for transportation to lunar surface on the Ares system or CEV). Point of contact: Tupper Hyde (NASA/GSFC)

The VSE architecture should include the capability for secondary payloads on both the Ares launch vehicles and the Orion space vehicles. These capabilities could include features such as an ESPA ring on the launch vehicles that could carry secondary payloads for deployment in near-lunar space, or the ESPA ring could form the structure for a secondary spacecraft—such as LCROSS—that could be deployed after the primary payload has been separated. Capabilities might also include secondary payloads for on-spacecraft autonomous instruments that do not require deployment. Orion should also have the ability to carry secondary payloads in an Apollo-like SIM that could be deployed in lunar orbit, as well as a payload bay that could accommodate remote sensing and in situ experiments with the necessary thermal, mechanical, power, and data handling interfaces.

In-space operations holds the potential for assembly, servicing, and deployment (trade studies), Point of contact: Harley Thronson (NASA/GSFC)

In a finding that was supported by presenters at this and other workshops, we found that very large aperture systems and spatial interferometers will be necessary to achieve many of the highest priority astrophysics goals. Such systems must operate at various locations in free space throughout the Earth-Moon system, such as at libration points and high-Earth and geosynchronous orbits. Capabilities to support these high-value systems will eventually become essential (e.g., assembly, service, repair, refuel). Such capabilities may be achieved by modest augmentations to NASA's Exploration Architecture, which will be operational during the same timeframe. Examples of enabling capabilities include robotic/telerobotics systems, advanced in-space EVA from Orion, and capable transportation such as the Ares system.

Large area lunar access to facilitate autonomous and/or human-assisted mobility (depending on trade studies). Points of contact: Joe Lazio (NRL) and Tom Murphy (UCSD)

Several high priority astrophysics programs are uniquely enabled by access to large areas of the lunar surface. Two concepts demonstrating this need are a large-area radio observatory located sufficiently far from human radio interference, and a widely dispersed retroreflector/transponder network to obtain increased accuracy for tests of general relativity. Both of these experiments/facilities could eventually require access to sites located hundreds to thousands of kilometers from a lunar base. Deployment of the assets could be done either autonomously or via astronauts.

Moderate Priority for astrophysics, may influence architecture

Minimize dust in the environment of small facilities (with optics, retro-reflectors).

High priority to astrophysics, will probably not influence Architecture

Enable high-bandwidth communication.

Evaluations and/or Trade Studies to Achieve Astrophysics Goals

Numbering is for ease of reference only and does not imply prioritization.

[1] Function of humans on lunar surface

Although lunar surface instruments have identified important science opportunities for astrophysics, these opportunities are either for small, mostly self-contained packages, or for facilities (e.g., long-wavelength radio interferometers, lunar-ranging targets) that do not require precision alignment or positioning. Therefore, while conveyance to the lunar surface is a requirement, the need of humans for emplacement, deployment, or operations may not be. Because of the possibility that general maintenance and servicing of such instrumentation may be uniquely enabled by hands-on access, a detailed assessment of the specific functionality of humans with respect to these opportunities should be done. This assessment can evaluate the viability of implementation plans that are entirely autonomous (or perhaps telerobotic), as well as to what extent such plans might tend to compromise the performance of these facilities. More broadly and in the context of the current exploration architecture, such an assessment could list the functional advantages by which a human agent could add value to any astrophysical installation on the lunar surface.

[2] Options for large-area lunar-surface emplacement

There are two astrophysical observations that require access to a large fraction of the lunar surface. First, a facility designed to observe the highly red-shifted hydrogen 21-cm line from the distant universe requires a significant amount of collecting area on the lunar far side—spread over at least tens of kilometers, and potentially even more. Current telescope designs envision a large number of individual elements (e.g., dipole antennas) that would need to be emplaced over this area. Second, sensitive tests of theories of gravity require on the Moon laser retroreflectors, transponders, or both. Optimal locations of these retroreflectors or transponders require wide spacing over the lunar surface at a variety of latitudes on the near side. An assessment is required of the manner or manners in which these elements (dipole antennae or retroreflector/transponders) would be emplaced across the desired area.

[3] Options for operations in free space

Capable operations in free space appear critical to achieve major goals for science, industry, and national security. Assessments and trade studies are necessary to more fully understand how these operations may enable multiple national priorities and provide a reliable basis for the design of elements of the lunar architecture. The assessment elements may include: (1) the function in space of astronauts and robotic partners; (2) technology investment strategies; (3) options for coordinated development with industry, other Government agencies, and international partners; (4) design options for block changes to the Orion/Ares systems; (5) cost estimates for possible modest augmentations to the exploration architecture; and (6) traceability of in-space systems to major national goals.

[4] Strategies to maximize the potential for a low-frequency observatory

The expected signal strength from highly red-shifted hydrogen is quite small (~10 mK), requiring dynamic ranges of at least 1 part in 10,000. Moreover, the signal is expected to be spread over a significant frequency (wavelength) range. In order to achieve such dynamic ranges and spectral access, a lunar telescope must be shielded from terrestrial, solar, and human-generated radio emissions. Generally, this requirement dictates a farside location. However, even on the far side, there are multiple options to both realize the telescope and preserve the radio frequency environment. Examples of potential tradeoffs include: (1) the degree of shielding and location on the far side, specifically with respect to how distant a long-wavelength observatory can be from a human outpost; (2) planning constraints for human and/or robotic sortie mission to farside exploration targets; and (3) the design of the communications infrastructure so as to maintain the radio frequency environment, particularly at low frequencies.

[5] Capabilities of the Ares system

Future major missions in space—both for science and national security—can be enabled by the capabilities of the proposed Ares 5 heavy-launch vehicle, specifically in regards to the mass and volume that can be delivered to priority locations throughout the Earth-Moon system. Assessment and trade studies to more fully understand how Ares 5 can enable multiple goals in space include: (1) detailed designs and performance estimates, including options for the fairing of alternative payloads (e.g., height, width, aspect ratio); (2) cost estimates, schedule, and milestones; (3) operation of the Ares V with other plausible systems operating during the same time period, such as the Orion or Ares I vehicles; and (4) recommendations for professional outreach to inform the science communities about the performance capabilities of the Ares vehicles.

Authors

Heidi B. Hammel, Space Science Institute, Boulder, Colorado, *hbh@alum.mit.edu*
John C. Mather, NASA Goddard Space Flight Center, Greenbelt, Maryland, *John.C.Mather@nasa.gov*

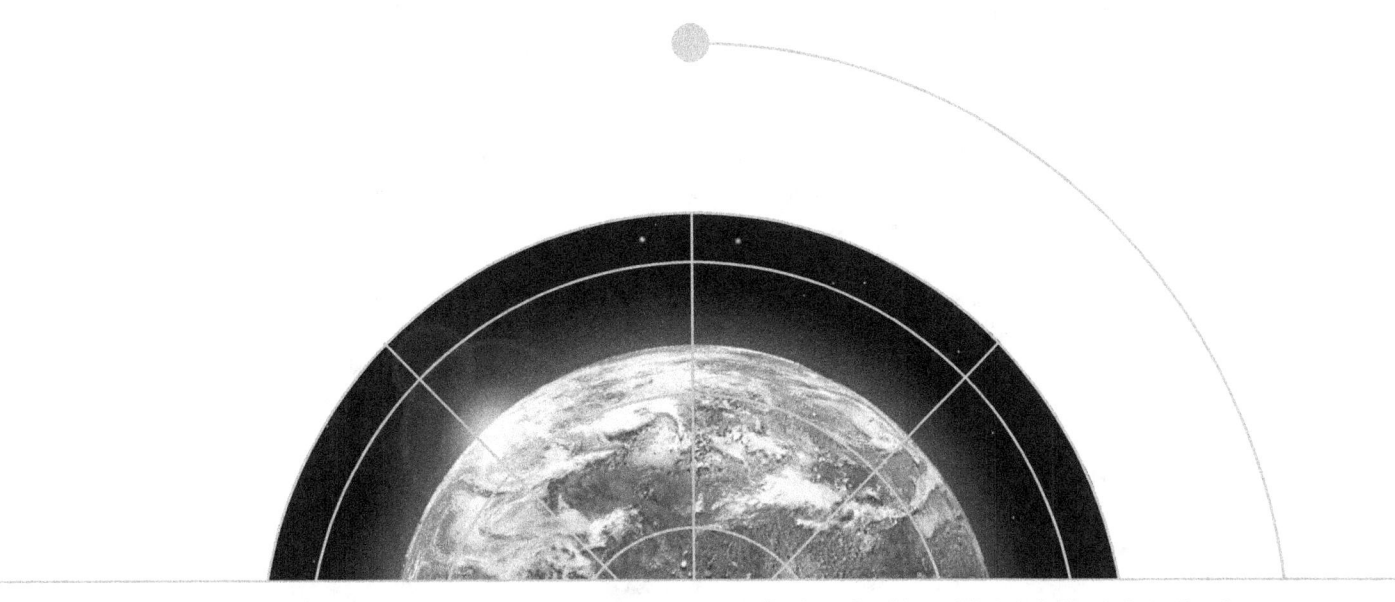

APPENDIX 2: WORKSHOP FINDINGS
EARTH SCIENCE SUBCOMMITTEE

The goal of NASA Earth science research is to understand the surface, atmospheric, and near-Earth space processes. To advance this understanding, we observe and model the Earth System to monitor its processes and discover the way changes occur. In so doing, we enable accurate prediction of changes and improve our understanding of the consequences of those changes for life on Earth. Much of the data needed for this research is currently collected by an array of LEO and GEO satellite-based instruments.

During the "Workshop on Science Associated with the Lunar Exploration Architecture" held in Tempe, Arizona, there were two overarching questions addressed by the ESS of the NASA Advisory Council and interested members of the science community:

1. What unique/complementary set of observations of the Earth can be made from the Moon that would significantly enhance data from LEO or GEO satellites?
2. Could those measurements be made from the proposed lunar South Pole outpost on the rim of Shackleton Crater?

These questions were addressed in a diverse set of talks presented in four scientific sessions: (1) A Lunar-based Earth Observatory, (2) Solid-Earth Science, (3) Atmospheric Composition and Climate, and (4) Sun-Earth Interactions.

The ESS concluded that a lunar Earth observatory would offer a unique, stable, and serviceable platform for global, continuous, full-spectrum views of the Earth to address a range of Earth science issues over time. In addition, such an observatory would provide instrument synergy among multiple LEO and GEO satellites for cooperative operations, enhanced calibration, and science. The proposed outpost location, while only offering limited views of the Earth, could still be initially useful for Earth science and instrument testing in the early stages of lunar exploration. However, the ESS endorsed a longer-term phased approach in which the future observatory would be located away from the outpost in order to provide the desired continuous Earth views while also mitigating the inevitable noise (e.g., radio, light, seismic, etc.) and dust problems associated with human activity, allowing for the collection of time-dependent data of atmospheric composition, ecosystem health, and hazard monitoring. This could be accomplished either from locations further to the north or south, at a higher elevation near the outpost, or from orbit at the Cislunar Lagrange Point (L1). The ESS also adopted the location criterion of unacceptable (if the Earth was in view less than 50 percent of the time), acceptable (if the Earth was in view more than 50 percent of the time), and desirable (if the Earth was viewable more than 90 percent of the time). The final location of such an observatory should be subject to careful analysis and study with the goal of a consistent architecture across instruments (e.g. communication links, compatible data formats, etc.) that would enable simplified instrument integration and expansion over time. Finally, a phased growth that begins with relatively simple instruments that are taken to the new outpost location, and develops into more complex instrumentation involving human or robotic sorties is recommended.

Regardless of the issues attached to the observatory location, the ESS recognized that there would be certain challenges and unique benefits by using the Moon as a remote sensing platform from which to observe the Earth. The rotation of the Earth as seen from the Moon would provide unprecedented temporal views of transient phenomena such as natural hazards, pollution, and climate. Furthermore, the Earth's orbital precession would allow observations of the polar regions—something not possible with GEO satellites. The Moon provides a stable and large platform for very unique remote sensing instruments—such as optical telescopes and long-baseline radar interferometers—that would be both accessible and serviceable, allowing the Earth to be monitored over the long term. In addition, many lunar-based remote sensing instruments can be more readily expanded and upgraded. However, the Moon is ~10 times further from Earth than GEO satellites, which makes acquiring data with useful spatial scales for smaller-scale processes more difficult. The Earth-Moon orbit also changes by ~5 percent through the year, making spatial resolution somewhat variable. Only limited views of the Earth would be possible depending on the time of day and day of the month/year. Finally, if instruments were located on the lunar surface, environmental factors (e.g., variable thermal conditions, those that may come from dust, etc.) would present challenges to instrument operations.

The concept of a lunar-based Earth observatory is highly compelling, but it must be planned so as to maximize the science return while not distracting from critically needed Earth science observations from other platforms. Certain Earth science observations can only be made well from LEO—such as high spatial resolution imaging and LIDAR—and these datasets should not be abandoned in the planning and implementation of a possible lunar-based Earth observatory. Furthermore, the ESS recommends that all future discussion and planning of Earth science return from the Moon be considered in light of the recently released NRC decadal survey, Earth Science and Applications from Space: National Imperatives for the Next Decade and Beyond (National Academies Press, 2007), while also recognizing that the decadal survey did not consider the options for observations from the Moon.

The following are the three primary concepts endorsed by the ESS and enabled by a future lunar-based Earth observatory. The list of science objectives (table 1) can be assessed, ranked, and placed within the overarching framework of these concepts.

1. **A dedicated Earth observatory at or on the Moon allows for global and continuous full-spectrum views of the Earth to address a range of Earth science issues.**

   The high temporal data frequency coupled with the ability to observe a given location for up to 12 hours enables detection and analysis of time-dependent atmospheric composition (i.e., global mapping of emissions, long-range transport of pollution plumes, greenhouse gases sources and sinks). This observational geometry makes new Earth and ecosystem monitoring abilities possible (i.e., volcanic eruptions, wildland fires, health and structure of vegetation, drought, and land degradation).

With climate change comes the critical need to observe changes in the cryosphere (i.e., ice shelf disintegration, sea ice change, snow cover cycles). A lunar platform also allows the Sun-Earth system to be observed simultaneously, providing data on the Earth's radiation balance and solar variability influence on climate. Finally, the numerous limb occultation opportunities over wavelengths from the visible (using stars) to the microwave (using GPS signals) to VHF (using communication signals) provide additional opportunities for observing the vertical structure of the Earth's atmosphere.

2. **The observatory provides a unique, stable, serviceable platform over the long term.**
   The location of the observatory will be critical to the amount, quality, and usefulness of the data returned. The location enabled most readily by, and with the lowest impact on, the proposed lunar architecture would be to place a series of Earth-observing instruments at the Cislunar L1 point (possibly being deployed by missions in transit to the lunar outpost location). This strategy has the benefits of being low cost (no down-mass carried to the surface and no sorties needed for surface instrument deployment), clean (no dust or thermal cycling contamination), and having unobstructed Earth views (no surface location constraints). Such an approach would still allow for longer-duration instrumentation and human or robotic serviceability in order to add, upgrade, or repair instruments. Despite those potential benefits, the ESS recommends a ground-based observatory as the first choice, allowing for much more growth and serviceability over time.

3. **The observatory would serve as a communications bridge across satellite platforms in other orbits (e.g., LEO, GEO, GPS).**
   A future lunar-based Earth observatory could also be used for enhanced calibration and science synergy with other orbital assets. For example, if a GEO/LEO satellite instrument with higher spatial resolution initially detected a thermal anomaly on a remote volcano, it could then task a targetable lunar-based IR instrument. The high temporal frequency (seconds) data from that instrument would be ideal for tracking the progression of the entire early stages of the eruption (i.e., the ash cloud migration). In the longer term, a lunar-based Synthetic Aperture Radar (SAR) whole-disk illuminator could also be used in conjunction with SAR receivers in LEO for surface deformation and cryosphere studies.

Overarching Earth Science Themes

The attending members of the ESS and interested/invited guests that participated in the Tempe workshop endeavored to assess the original scientific objectives in light of the LAT's rankings, determine how these objectives/rankings would be impacted by the proposed outpost location, understand the possible science that could be accomplished from an Earth observatory on the Moon, and adopt recommendations for the Council and the LAT. Furthermore, in a time of shrinking budgets, all lunar discussions were tempered by the recently released NRC decadal survey that called for a substantial increase in both Earth science funding and new missions. The ESS also considered the objections of many in the Earth science community to the overall concept of locating Earth-observing instrumentation on the Moon who see such a deployment as a diversion of future limited resources (made even more poignant by the recommendations put forth in the decadal survey) away from LEO or GEO.

However, regardless of the decadal survey's impact, the primary task of the subcommittee was to critically assess the original science objectives for a lunar outpost in light of the low rankings given to most of those objectives by the LAT with respect to the capability to accomplish them within the constraints of the notional polar-outpost architecture. A second and related task was to assess how Earth science objectives could be better achieved at a location different from Shackleton Crater. Through invited presentations, posters, and general discussions, the subcommittee worked to assess the capabilities of the lunar architecture to achieve those objectives and determined both the intrinsic scientific value and our best understanding of how well the objectives meshed with the proposed architecture.

The results are summarized below in four main sections. We first present a framework of three fundamental tenets within which the subsequent detailed assessment of Earth science objectives should be framed. We then describe a phasing strategy and the Earth science capabilities enabled by the VSE. Following that are two sections: (1) Required Studies/Factors Needed to Achieve Earth Science Goals, which describes the key research studies needed prior to further development of the lunar architecture as it relates to specific Earth science objectives; and (2) Emerging Technologies for Earth Science, which identifies new and innovative technology developments that would be important for implementation of the overarching science themes. Lastly, we summarize the public outreach that could stem from this observatory.

Detailed Assessment of Earth Science Objectives

The original list of science objectives was crafted by the ESS at the September 2006 subcommittee meeting (table 1). We present these objectives with several new levels of assessment that were based on the expertise at the Tempe workshop and the recommendation that a future Earth observatory be located away from the proposed outpost site in order to maximize Earth viewing. In addition, the criteria below must be considered prior to implementation of the science objectives or revision of the lunar architecture.

1. **There are worthwhile and important Earth science opportunities enabled by a lunar outpost.**
   There was an assumption by many in the Earth science community (as well as the LAT) that Earth observations from the Moon would require very large telescopes (>> 1 meters) and therefore would not be feasible. This is factually not true, and several presentations were made showing the potential science return using telescopes as small as ~ 0.3 meters. These relatively modest instruments have already been flight tested (e.g., HiRISE—High Resolution Imaging Science Experiment—on the MRO—Mars Reconnaissance Orbiter). Furthermore, significant technological advances are expected in the next 10 to 20 years that may further reduce the size, mass, and complexity of such telescopes.

   The viewing geometry of the Earth from the Moon will be both a benefit and a hindrance depending on the type of science observation needed. The spin and precession of the Earth enable instruments on or at the Moon to view constantly changing conditions and track objects with very high temporal frequency during the viewing intervals. However, the phases of the Earth, the variable Earth-Moon distance, and the inability to observe certain locations continuously for long periods will hinder some optical remote sensing objectives.

Three examples of fundamental science (in no particular order) made possible only from a lunar viewing position are: (1) the collection of "whole Earth" spectral data as a calibration source for future terrestrial planet finder missions, (2) the ability to track temporally variable atmospheric pollution and volcanic plumes, and (3) the rapid response to natural disasters in coordination with LEO and GEO assets.

2. **The VSE should be planned so as to accommodate capabilities that will enable Earth science.**
Earth science observations will become increasingly critical in the coming decades with accelerating climate change and the need to monitor and, if possible, forecast natural disasters. Furthermore, the psychological impact of seeing our home world in the vastness of space cannot be underestimated. Therefore, we feel it absolutely critical that worthwhile Earth science be conducted from the Moon above—beyond the occasional astronaut photograph. Without significant mobility, Earth-observing science is seriously limited at the notional Shackleton Crater outpost location; therefore, cost-effective alternatives should be considered.

3. **Any lunar-enabled science can and should be evaluated and prioritized within the Earth science community by the decadal survey process.**
The Earth Science Directorate will need to prioritize and commit funds to activities and missions outlined in the new decadal survey. Our assessments should not be considered in anyway to replace or supersede the decadal survey process. However, we also recognize that the recent Earth science decadal survey was conducted without any consideration of future lunar assets, which would be deployed as exploration continues. The assessments below include, in addition to intrinsic science, the manner in which the science may mesh with the future lunar architecture.

Table 1
ESS Science Objectives and Assessment for a Lunar-Based Earth Observatory

Assessment Column: Colors/numbers synchronize with the original LAT assessment. Each number signifies a different assessment level from 1 (easily doable) to 5 (not doable at all) within the notional lunar architecture. Note that these values do not rank the objective's science potential, but rather are based on how easily the objective can be met within the proposed architecture. Three assessment levels are given:
- 1st value: original LAT objective-to-architecture rating
- 2nd value: modified LAT objective-to-architecture rating, reassessed by ESS)
- 3rd value: modified LAT objective-to-architecture rating, enabled by an alternative Earth-viewing location

Science Ranking Column: The ranking for each objective is dominated primarily by the expected science return and assumes an optimal Earth viewing location. A minor component of the ranking score is also the mission phasing timeline (table 2) and the infrastructure required to implement the particular objective (see the "Earth Science Capabilities Enabled by the VSE" section).

- [A] = highest science priority and low impact on the current lunar architecture
- [B] = high science priority and moderate-high impact on the current lunar architecture
- [C] = medium-high science priority and high impact on the current lunar architecture

Code	Short Title	LAT/ESS Assessment	Comments	Science Ranking
mEO1	Monitor the Earth's Magnetosphere	[4] / [4] / [4]	Ground or L1-based instruments can be used to observe the Earth's magnetosphere in order to develop predictive and mitigation capabilities for magnetosphere-driven events (in conjunction with the Heliophysics Subcommittee [HPS]). This is best-driven by HPS, and without feedback from them, the original ranking was not changed.	[B]
mEO2	Create Topography, Altimetry, and Tomography Maps	[5] / [5] / [4]	Using SAR and multi-baseline Interferometric Synthetic Aperture Radar (INSAR) from the lunar surface either with co-located receivers or with ones in LEO would provide high temporal resolution, full Earth deformation and topographic mapping. This is a high priority for the Earth science community. However, the need for a nearside location, possibly nuclear power, and major infrastructure has kept this objective ranked low.	[B]
mEO3	Characterize the Earth's Atmospheric Composition and Dynamics	[4] / [2] / [1]	A hyperspectral sensor ranging from the UV to the TIR—much like the current Ozone Monitoring Instrument (OMI), Transition Edge Sensors (TES), and Atmospheric Infrared Sounder (AIRS)—coupled with the near-constant limb profiles of Earth could be used to map SO2, O3, CO, CH4, NO2, HNO3, plumes, and sources/sinks. Full Earth views are critical, but the telescope could be as small as 30-50 cm (hence, the improved [2] / [1] ranking).	[A]
mEO4	Monitor the Sun-Earth System	[4] / [2] / [1]	Understanding the effect of solar variability on Earth's atmospheric composition and climate would be uniquely enabled from an instrument at or on the lunar surface. Full Earth and Sun views are critical, but the telescope could be as small as 30-50 cm (hence, the improved [2] / [1] ranking).	[A]
mEO5	Determine the Earth's Bidirectional Reflectance Distribution Function (BRDF)	[4] / [2] / [1]	Hyperspectral observations at multiple incidence, emission, and phase angles can provide more precise radiative balance calculations than currently available from Earth-orbiting satellites for climate studies. Full Earth views are critical, but the telescope could be as small as 30-50 cm (hence, the improved [2] / [1] ranking).	[B]
mEO6	Measure the Earth's Ocean Color	[5] / [5] / [4]	Although 70% of the Earth's surface is covered by water, and ocean observations should be numerous from the Moon, feedback thus far from the ocean community has been pessimistic. They think that meaningful science can be done only from LEO. Therefore, we have kept this objective's ranking low, but we continue to keep it in the table pending a broader examination by the ocean science community.	[C]

mEO7	Map the Surface Composition of the Earth	[4] / [2] / [1]	A multispectral sensor ranging from the UV to the TIR (much like MODIS) could fulfill several objectives on this list (mEO3, mEO5 and possibly mEO4, mEO12). Full Earth views are critical, but the telescope could be as small as 30-50 cm (hence, the improved [2] / [1] ranking).	[A]
mEO8	Measure the Historical Solar Constant	[1] / [2] / [1]	Information on solar variability over the past centuries through borehole thermal conductivity measurements could be recovered initially with smaller boreholes at the outpost site. This would be expanded as drilling technology is improved on the lunar surface and sorties are made to the near side.	[A]
mEO9	Observe the Earth's Ice Surfaces Over Time	[5] / [5] / [4]	To understand how ice cover is impacted by climate change, the extent and volume must be measured. Using SAR from the lunar surface would provide high temporal resolution ice mapping covering the poles. This is a high priority for the Earth science community. However, the need for a nearside location, possibly nuclear power, and major infrastructure has kept this objective ranked low (see mEO2).	[B]
mEO10	Monitor Earth's "Hot Spots"	[5] / [2] / [1]	Thermally elevated features (volcanic, fire, and anthropogenic activity) could be monitored with high temporal frequency (and in conjunction with LEO and GEO satellites). This instrument could be phased in from a simple radiometer to a multispectral sensor. Full Earth views are critical, but telescope could be as small as 30-50 cm (hence, the improved [2] / [1] ranking).	[A]
mEO11	Calibrate Earthshine	[1] / [1] / [1]	The objective is to measure true Earth albedo (and cloud amount, etc.) from the Moon, and calibrate these results with current and past Earth-based Earthshine measurements. This could be accomplished by using the other proposed instruments/science listed here (mEO3, mEO4, mEO5, mEO12), but does not need long term, full-Earth views (hence, the [1] / [1] ranking).	[B]
mEO12	Observe Lightning on the Earth	[[5] / [2] / [1]	A narrow band (0.774 μm) detector with 10 km spatial resolution for detection and mapping of lightning for climatology, monitoring, and hazard mitigation (tornadoes, severe storms, etc.). Full Earth views are critical, but the telescope could be as small as 50-100 cm (hence, the improved [2] / [1] ranking).	[A]

In order to achieve the maximum return on future Earth science from the Moon and best integrate with the final lunar architecture, the ESS recommends a phased approach to instrumentation. This phasing would begin with relatively simple instruments deployed into either an L1 orbit or at the surface by humans, eventually extending to more complex instruments requiring significantly more infrastructure. Therefore, we have factored this expansion into the ranking column (table 1) and urge the LAT to consider this approach during future architecture planning. In table 2, short-term phasing would occur during the early years (2020-2025) of the lunar outpost. Instruments would be modest cameras and/or spectrometers either placed in L1 orbit or set up and tested on the lunar surface near the outpost. If the latter, Earth observations would be limited, but initial instrument testing in conjunction with some science return would still be worthwhile. Midterm phasing (2025-2030) would involve sorties away from the outpost and begin with the establishment of the permanent Earth observatory structure at the chosen location for optimal Earth viewing. A high scientific return is expected from this phase. Alternatively, if the observatory is to be completely orbital, this phase would see enhancements of the existing instrument complement. By the end of this phase, the Earth observatory instrument suite (for the [A] and [B] rankings) would be complete and regular, and long-term Earth observations would be underway. Finally, the long-term phasing (2030 and beyond) would include the addition of significant infrastructure and power sources, especially for active instruments, and longer-term sorties to the other parts of the lunar surface. Active remote sensing such as the lunar-based SAR could come online in this phase.

Table 2
Proposed ESS Mission Phasing Timeline and Examples

Phase	Years	Comments/Examples
short-term	2020-2025	Ground or L1-based instruments would be used to observe the Earth's magnetosphere to develop predictive and mitigation capabilities for magnetosphere-driven events (in conjunction with HPS). This is best-driven by HPS, and without feedback from them, the original ranking was not changed.
mEO2	2025-2030	More complex and longer duration instrumentation would be deployed either in Cislunar L1 orbit or on the surface at the permanent observatory location. This period would serve as the transition to long-term monitoring of critical Earth science variables. New instrumentation and upgrades expected throughout. High science return expected. Examples: enhancements (e.g., larger foreoptics, new spectrometers, etc.) of existing complement; test drill holes (2-10 m) for thermal conductivity measurements (i.e., mEO8)
long-term	> 2030	This phase requires very complex infrastructure (nuclear power sources, deep-drilling capability), and long distance (equatorial near side) sorties. Examples: microwave (SAR) illumination of entire Earth disk; LIDAR measurements (atmospheric composition, vegetation structure, ice deformation); and deep-drilling (100 m) for heat flow/solar constant

Earth Science Capabilities Enabled by the VSE

Examples of some of the science enabled by observing the Earth from the Moon are described below. These data would complement Earth orbital observations and provide well characterized observations for long-term trends. Most importantly, the lunar platform would enable new observations and new technologies not possible from LEO or GEO. The following section more fully describes the expected science return from a lunar-based Earth observatory and summarizes the information presented by many of the invited speakers for each of the ESS objectives (table 1). Within each category, no prioritization is implied.

[A.] **Highest priority Earth science that may influence lunar architecture planning.**
   Objectives that are fully or partially enabled in the short term and are of the highest science priority include: mEO3, mEO4, mEO7, mEO8, mEO10 and mEO12.

Example 1: Rapid response time

1. P. Christensen (Arizona State University) summarized the concept of a modest imager having a 0.3 meter aperture with a 0.2° IFOV and a 2,048 pixel array (similar to the HiRISE Camera Mars Reconnaissance Orbiter) that would provide 0.5 kilometer/pixel (VNIR); 1-2 kilometer/pixel (SWIR); and 10 kilometer/pixel (LWIR). Such an imager would only cover a 1,000 kilometer by 1,000 kilometer field of view during a given scan. However, if the sensor was made to be pointable, it could be integrated into a sensor web concept with LEO and GEO satellites to quickly target any given location on Earth. This instrument would be part of an initial instrument suite within the Earth observatory, and be upgradeable over time to incorporate new technologies, operate in research mode, and provide real-time link between GEO and LEO observations.

2. J. West (NASA-Jet Propulsion Laboratory [JPL]) discussed the potential of leveraging the unique advantages of the Earth-Moon (Cislunar) L1 vantage point for the placement of Earth-observing satellites. This location offers continuous staring opportunities at the Earth (and back at the Moon). The advantages of the L1 Earth observatory include potentially lower cost (no down-mass to the lunar surface required), no contamination (e.g., surface dust) potential, and unobstructed whole-Earth views. The cost of mission-specific upgrades and maintenance will need to be evaluated. This kind of orbital observatory could be implemented using small, instrumented, autonomous mini-satellites deployed by the astronauts from the crew transfer vehicle on the way to the Moon.

3. M. Ramsey (University of Pittsburgh) summarized the current near-real-time monitoring of thermal anomalies (hot spots) using a sensor web concept between GEO satellites and higher resolution LEO instruments. That program exists in the northern Pacific region and uses moderate to low spatial resolution TIR instruments—e.g., Advanced Very High Resolution Radiometer (AVHRR), Geostationary Operational Environmental Satellites (GOES), and MODIS)—for the initial detection and triggering of high spatial resolution TIR instruments—e.g., Advanced Spaceborne Thermal Emission and Reflection Radiometer (ASTER) and Enhanced Thematic Mapper Plus (ETM+). The data collection is on the scale of minutes and directly applicable to real-time hazard tracking (e.g., volcanic plumes). In the future, an initial detection by LEO or GEO could trigger the lunar TIR instrument operating in the 3-12 μm region. Most importantly, that instrument could observe the volcanic eruption continuously at very high temporal resolution. For large eruptions, the data would be unprecedented, capturing the initial stages and progress of the aerosol/gas plumes. Similar opportunities exist for observations of other disasters.

Example 2: Unique viewing geometry

1. S. Goodman (NASA-MSFC) presented the possibility of performing observations of lightning on Earth from the lunar surface. The detection and global monitoring of lightning has important implication for severe weather hazards, global production of nitrogen oxide (NOx), and coupling with the magnetosphere. The high-speed (500 frames per second) sensor would be centered at 0.774 micrometer and provide 10 kilometer spatial resolution.

2. J. Herman (NASA-GSFC) introduced the concept of simultaneous measurements from the Moon of the Sun, its solar ejections, and their effects on Earth. The data would allow a better understanding of the processes and interactions that determine the composition of the Earth's whole atmosphere, including the connections to solar activity. The data could also be used to map atmospheric species concentrations (greenhouse gases, aerosols, ozone) and provide real-time space weather data for predictive modeling of the space environment.

Example 3: Earth science on the lunar surface

1. K. Steffen (University of Colorado) presented the potential of measuring the solar constant (TSI) on the lunar surface. TSI is one of the most important climatic factors, and has influenced the Earth's climate in the past. However, retrieving detailed measurements of the past TSI is not possible on Earth. Unlike Earth, the lunar surface is in a state of radiative equilibrium with the Sun; therefore, its surface temperature is determined by TSI directly. By measuring the temperature profile in lunar boreholes, the past TSI can be recovered. The ideal site for these measurements would be near the lunar equator (large absolute flux and better resolution for TSI) and a 100-meter borehole would resolve data back to 1600 A.D.

[B.] High-priority Earth science that may or may not influence lunar architecture planning. Objectives that are fully or partially enabled in the midterm and are of high science priority include mEO1, mEO5, and mEO11.

Example 1: Unique viewing geometry

1. M. Turnbull (Space Telescope Science Institute (STScI)/Carnegie) focused on the unique viewing of the Earth from the Moon to ask the question: Are there any astrophysics projects that are uniquely enabled by the lunar platform? The ability to collect whole-Earth, full-spectrum, spatially resolved views would provide a unique calibration dataset for future terrestrial planet finder missions. The detailed data from the Moon of the variable Earth would be important for identifying and characterizing habitable worlds around nearby stars (the spatially unresolved case).

2. J. Mustard (Brown University) focused on land surface monitoring from the Moon and its unique observation conditions (changing incidence and emergence angles and the 28-day repeat of illumination conditions). In particular, the lunar observatory would provide an important measure of the BRDF. The BRDF is capable of retrieving certain properties, such as ecosystem structure, and collection from the Moon would more completely sample (e.g. near 0 phase) the full BRDF for science applications. In addition, plant phenology (timing and magnitude of ecosystem processes indicated by greenness) could be measured as a function of time.

3. N. Loeb (NASA-Langley Research Center [LRC]) compared current monitoring of the Earth's albedo from LEO satellites, such as the Cloud and the Earth's Radiant Energy System (CERES) instrument, to what might be possible from the Moon. Specifically, he focused on two

questions: (1) What are the climate accuracy requirements for monitoring the Earth's albedo? and (2) Can the Earthshine approach (i.e., from the Moon) satisfy these climate accuracy requirements? The detailed modeling presented initially indicates that albedo measurements of the Earth from the Moon are unlikely to achieve 0.3 Wm-2/decade stability requirement needed for precise climate science. However, this measurement could still be an important validation for future LEO data, and more modeling is needed before this Earth Science objective (mEO11) is abandoned.

4. A. Ruzmaikin (NASA-JPL) also examined the possibility of measuring the Earth's broadband albedo (0.3 to 3 micrometers) from the Moon for the purposes of better climate modeling. Deviations in albedo can be caused by many factors (e.g., seasons, latitude, clouds, etc.), which can propagate errors into climate models. The benefits of a lunar-based albedo measurement were found to be homogeneous longitude sampling, high temporal (hours) and spatial resolution (10 kilometers), observation of the polar regions, observation of the diurnal albedo cycle, and a potentially much longer lifetime than any LEO satellite can provide.

Example 2: Active remote sensing from the Moon

1. K. Sarabandi (University of Michigan) presented the intriguing potential of conducting large baseline synthetic aperture radar interferometry of the Earth from the Moon. The objective would be to create solid Earth, topography, altimetry, three-dimensional tomography, and vegetation maps. SAR images would be formed using the relative motion of the Earth with respect to the Moon by having multiple antennas to form a microwave interferometer with a long baseline and extreme stability. This configuration also allows for multi-static operation in conjunction with relatively inexpensive SAR receivers in LEO. Although the implementation of this science objective would require significant infrastructure, the instrumentation would provide a whole-disk illumination of the Earth in the microwave band allowing continuous, all-weather observations of the planet.

Required Studies and/or Factors Needed to Achieve Earth Science Goals

Certain studies must be carried out and other factors considered in the short-term prior to any continued lunar architecture planning. The following list highlights these subjects. Numbering is for ease of reference only; no prioritization is implied.

1. **Options for lunar-surface emplacement**
   If a future Earth observatory is to be located on the lunar surface, engineering studies must be conducted to determine the best strategy for maximizing the Earth observation potential. The study of possible locations should include sortie locations within easy reach of the lunar outpost. These could include a lower-elevation site either further north or south, or a higher elevation site (e.g., Mt. Malapert) in closer proximity to the outpost. Both would possibly require new logistical and infrastructure considerations for the current lunar architecture. The location must at minimum meet the

acceptable criterion (Earth observed > 50 percent of the time) and ideally would attain the desirable criterion (Earth observed > 90 percent of the time).

2. **Options for operations in free space**

    Because of the limited options and cost associated with a lunar surface Earth observatory, the second option would be to have instruments placed at the Cislunar (L1) point in order to provide full-Earth views and achieve the major goals for science. Assessments and trade studies are necessary to understand how these operations may be enabled within the lunar architecture. The assessment elements may include: (1) the capacity of the Orion/Ares systems to carry and deploy small satellites prior to arrival at the Moon; (2) technology, maintenance, and enhancement strategies and tradeoffs compared to surface-based instruments; (3) options for coordinated development with other partners; and (4) cost estimates for possible modest augmentations to the exploration architecture.

3. **More formal modeling of sensor design needed and data quality expected in order to address the science objectives**

    More precise and formalized engineering studies must be carried out in order to constrain both the common architecture desired in a future Earth observatory and the specific sensor designs (i.e., power requirements, size, mass, orbital vs. landed). The sensor designs should consider both the Earth science objectives (table 1) and the proposed mission phasing (table 2), and have detailed input from scientists working in these fields. The design criteria should, if possible, carry forward both space-based and surface-based options with specific tradeoffs for each. For example, the complications of the lunar thermal environment and those that hypothetically may come from dust should be considered especially for larger optical telescopes. The universal architecture for a permanent lunar-based observatory must also be made in conjunction with the final lunar architecture and able to easily integrate.

4. **Involve the Earth science community**

    The ESS should organize and plan an Earth Science from the Moon workshop (similar to that held by the Astrophysics Subcommittee in November 2006). This should be carried out within one year of the Tempe workshop and involve a wide array of Earth scientists, engineers, the LAT, and representatives from ESD/ESMD. The current science objectives should be revisited and finalized at that time. Ideally, initial mission trade studies (see number 3) would have been conducted so that they can be presented at this time. Furthermore, the LAT should present the feasibility of sortie locations (ground or orbital) for the Earth observatory (see numbers 1 and 2).

5. **Function of humans and instrumentation on the lunar surface**

    Opportunities have been identified for Earth science from lunar surface instruments that must be based away from the outpost location. In this context, conveyance to the lunar surface and deployment to the observation site is a requirement; however, humans may not be needed for these processes. If general maintenance and servicing of such instrumentation is required over time, it may be enabled by astronaut access (or perhaps telerobotic operations). Therefore, a detailed assessment of the specific functionality of humans with respect to these opportunities should be done. This assessment would evaluate the viability of the Earth science plans for instrument deployment and the functional advantages by which an astronaut could add value to any installation on the lunar surface.

Emerging Technologies for Lunar-Based Earth Science

During the discussion of the possible Earth science enabled by the lunar architecture, several new and innovative technologies were all briefly mentioned. These were primarily focused on imaging, orbital, and power technologies, and they could all significantly improve the data return from the Moon. The concepts listed below should be considered in future planning. Numbering is for ease of reference only; no prioritization is implied.

1. **Active pixel sensor (APS) for effective whole-Earth imaging with reduced data rate**
   The APS is an imaging device similar to the charge-coupled device (CCD). But in contrast to CCD, each APS pixel contains a photodetector and is connected to a transistor reset and readout circuit. This allows selection of the whole set of image pixels or only a subset of pixels for readout, thus focusing on interesting parts of the image with the reduced data rate. APS consumes far less power than CCD, has less image lag, and is cheaper to fabricate. The larger arrays and lower power requirements could allow the whole Earth disk to be imaged at moderately-high spatial and spectral resolutions.

2. **Spectrally resolved pixels for large imaging arrays**
   In this CCD, which can be used for spectral imaging, each pixel is actually a microspectrometer acting simultaneously and independently of other pixels. As a result, spectral imaging acquires a cube whose appellate signifies the two spatial dimensions of a two-dimensional sample (x and y), and the third is the wavelength dimension. Practically, it must be combined with a monochromator that diverts light of different wavelengths onto different pixels. This CCD simultaneously collects photons in a broad wavelength range, enabling to measure an entire spectrum in a very short time.

3. **Solar electric propulsion, nuclear electric propulsion, or solar sail allowing for a "pole-sitting" observatory**
   Positioning a long-lived satellite far below the lunar south pole would require propulsion and station keeping technologies. This would serve several potential key applications. Most importantly, it would enable real-time, wide regional observation of the outpost and its surroundings, as well as simultaneous views of the Sun, Earth, and Moon from different angles. It could function as a continuous communications node between the Earth and the Moon and/or between the outpost and lunar sortie missions. Depending on the instrumentation on such a satellite, it could also serve as a stable remote sensing platform for observations of the lunar southern hemisphere. Other uses could include solar-wind monitoring and a relay for future deep-space missions.

Outreach and Public Impact

The psychological impact of seeing Earth from space should not be underestimated. Images from the Apollo and Galileo missions provide a global view of our home planet not seen from either LEO or GEO based instruments. However, we must expand beyond the occasional photograph of the Earth to a more systematic and synoptic set of measurements that can only be realized and enabled by the VSE.

We propose that it would be a serious flaw in the VSE and the proposed lunar architecture if an outpost location is chosen with little to no opportunity to perform quantitative Earth science, which can then be used to inspire the public. To highlight this concept, we include a quote from the opening statement of the former Chairman of the Committee on Science for the House of Representatives, Sherwood Boehlert (R-NY) on April 28, 2005: "The Earth science program doesn't exist as some secondary adjunct of the exploration program ... there's no reason that NASA can't robustly carry out the President's Vision for Space Exploration while conducting vital Earth science research."

The Earth science members participating in the workshop were asked to craft opportunities for public outreach expected from our proposed Earth observatory on the Moon. The dominant themes are presented here. Numbering is for ease of reference only; no prioritization is implied.

1. **The "Blue Planet Webcam"**
   In the process of collecting visible and infrared spectroscopic data for the proposed science objectives (e.g., mEO3, mEO7, mEO10), regular visible images of the Earth would be generated. These real-time, whole Earth views would be an amazing educational resource that could be visualized in an online environment along the lines of "Google Earth."
2. **Building the Lunar-Based Earth Observatory**
   If an actual observatory is built on the lunar surface to observe Earth, the overarching imagery of an "observatory on a hill" is expected to be very compelling. This iconic view of what an observatory is on Earth (e.g., the telescope under the white dome on the mountain) would be duplicated on the Moon in order to look back at Earth. The data collected from the instruments that comprise the Earth observatory will be used for long-term synoptic environmental monitoring, which will become increasingly important with accelerated climate change. Furthermore, a future terrestrial planet finder mission will be able to use these data as a critical calibration source.
3. **Taking the Pulse of Earth from the Moon**
   Related to the previous two points is the very reason these data would be collected: to monitor the Earth and acquire critically needed measurements from which to model trends in the atmosphere, lithosphere, cryosphere, hydrosphere, and biosphere. Tracking climate variability, air pollution sources and transport, natural hazards (e.g., extreme weather, volcanic plumes, hurricanes, lightning), seasonal and secular variations in polar ice, and vegetation health (e.g., spring greening) were all identified in the workshop as feasible and important data that could be collected from the Moon. Such data would be important both for public consumption and useful for many different NASA projects.

Authors

Michael Ramsey, Department of Geology and Planetary Science, University of Pittsburgh, Pittsburgh, Pennsylvania, *ramsey@ivis.eps.pitt.edu*

Jean-Bernard Minster, Scripps Institution of Oceanography, University of California, Institute of Geophysics and Planetary Physics, La Jolla, California, *jbminster@ucsd.edu*

Kamal Sarabandi, Engineering and Computer Science Department, The University of Michigan, Ann Arbor, Michigan, *saraband@eecs.umich.edu*

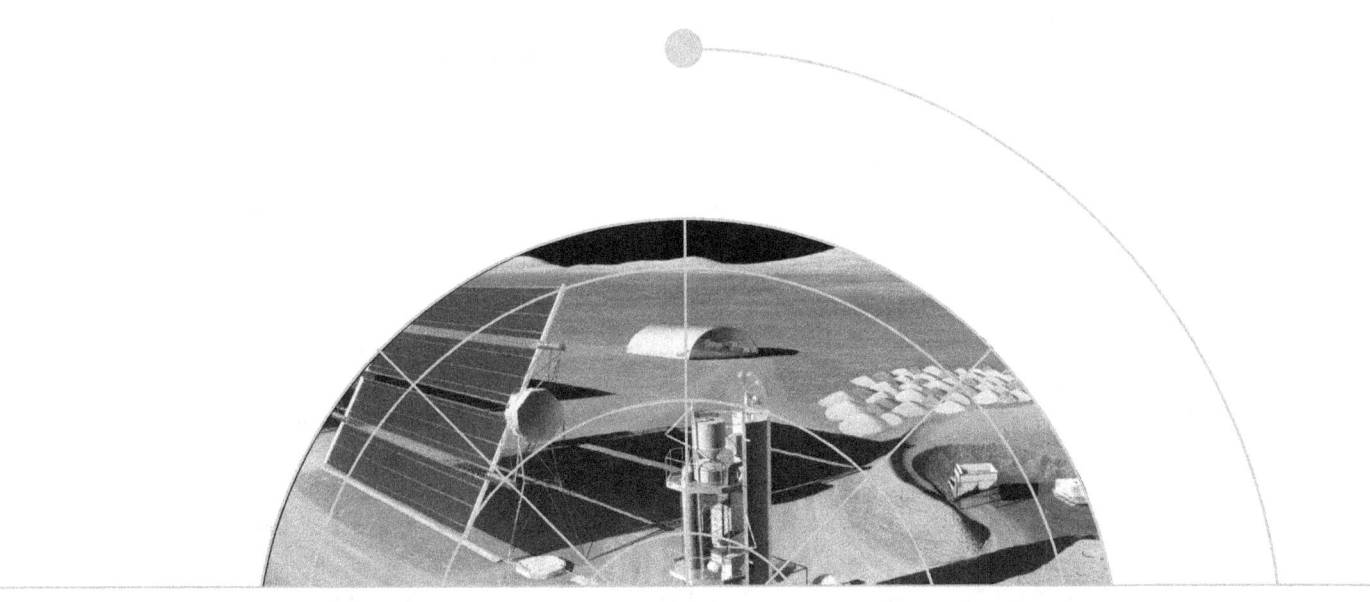

APPENDIX 3: WORKSHOP FINDINGS
HELIOPHYSICS SUBCOMMITTEE

Heliophysics Science and the Moon Synthesis

Members of the NAC Heliophysics Subcommittee and interested members of the community considered at length several space science topics drawn from community input over the previous eight months by the subcommittee's Heliophysics Science and the Moon subpanel. During the deliberations, it was apparent that the architecture potentially available by NASA's return to the Moon presents interesting and exciting new opportunities to extend scientific progress in ways that have not been previously available or considered. The synthesis of these deliberations is contained in this report. A separate report, "Heliophysics Science and the Moon," from the Council's Heliophysics Subcommittee—released summer 2007—provides more detail on the potential solar and space physics science for lunar exploration.

Since the inception of the space program with Explorer 1 and continuing through to the present space weather missions, scientists in the heliophysics community have worked to develop a detailed understanding of the connected Sun-Earth-Moon system. The Moon is immersed in a plasma environment—the local cosmos—that is magnetized. These fields play an essential role in organizing the environment. It is the twisting and folding of the various interacting magnetic fields—of the Earth, of the Sun, and of the Moon itself—that regulate the local environment of the Moon and, thus, the environment that will be experienced by human explorers. By working to understand this environment and, ultimately, to predict the variations likely to occur from day to day and region to region, it is widely believed that the productivity of future lunar robotic and manned missions can be significantly enhanced.

The heliophysics science topics related to lunar exploration are grouped in four themes: (1) Heliophysics Science of the Moon—investigating fundamental space plasma processes using the Moon and its environment as a natural laboratory; (2) Space Weather: Safeguarding the Journey—understanding the drivers and dominant mechanisms of the lunar radiation and plasma-dust environment that affect the health and productivity of human and robotic explorers; (3) The Moon as a Historical Record—seeking knowledge of the history and evolution of the Sun and solar system as captured in the lunar soil; and (4) The Moon as a Heliophysics Science Platform—exploring possibilities of establishing remote sensing capability on the lunar surface to probe geospace, the Sun, and the heliosphere.

Subcommittee Workshop Conclusions

The Heliophysics Subcommittee discussed various opportunities for science related to lunar exploration. Several issues were raised during the week. Of those, the following were deemed crosscutting and/or important to heliophysics science and the Moon.
- For several heliophysics science opportunities, drop-off satellites or early robotic operations are optimal.
- Lunar science assessments formulated at this workshop are deemed to be valuable input to the next NRC Decadal Survey for Solar and Space Physics and NASA Heliophysics Science Roadmap. NASA SMD has a well-validated process for establishing science priorities within their resource

allocations. Once complete, the lunar science opportunities information should enter into this process in the same manner as other SMD pre-planning activities.
- Future evaluations of these science objectives must assess the cost effectiveness of these lunar site implementations versus implementations that utilize robotic/unmanned missions around the Moon or elsewhere.
- For full mission success, many of these science objectives will necessarily require involvement of a scientist-astronaut as an integral part of the science experiment.

Detailed Assessment of LAT Objectives Associated with Heliophysics

The subcommittee assessed each of the objectives identified by the NASA Lunar Architecture Team as being related to heliophysics. Full assessments for all concepts will be contained in the Heliophysics Subcommittee report on Heliophysics Science at the Moon.

The assessment—summarized in the table below—was performed according to the criteria below. Please note that the Objective-to-Architecture rating is provided in the fifth column.
- **High:** Science is of high value and achievable within the architecture, or the importance to lunar operations is deemed high.
- **Medium:** Science is of secondary value and achievable within the architecture, or the objective is deemed important to lunar operations.
- **Low:** Little or no science return, or the likelihood of achieving the objective within notional architecture is low.

The objective-to-architecture rating in this table follows the same scheme as presented by the Science Capability Focus Element (SCFE) of the NASA Lunar Architecture I development as summarized in the following table:

[1] Objective can be substantially accomplished by 2025 within the current architecture assuming the priority and funding are allocated.
[2] Objective will very likely take longer than the 2025 time horizon to accomplish, but could be accomplished in an outpost-based architecture.
[3] Some substantial part of the objective can be accomplished within the current architecture by 2025.
[4] Objective can be accomplished with a combination of outpost-based science and robotic sorties.
[5] Objective can really only be accomplished through the addition of human sorties, selection of a different site for the outpost, or the addition of some other capability such as long-range mobility, to the current architecture.

Objective ID#	Title	Assessment	Comments	Suitability to Single Site Architecture
mHEO3	Study the dynamics of the magnetotail as it crosses the Moon's orbit to learn about the development and transport of plasmoids.	High	The dynamic behavior of the distant magnetotail, where a substantial fraction of the total energy coupled into the magnetosphere from the solar wind is stored, is not understood. It is different from the near-Earth, with quasi-continuous, physically different magnetic reconnection. The Moon is a unique location for studying the deep magnetotail, allowing diagnostics of the magnetic field topology and convection velocity by observations of lunar shadowing of ambient electrons.	Requires an orbital mission, perhaps as a dropoff satellite. Objective-to-architecture rating: [1]
mHEO4	Study the impact of the Moon on the surrounding plasma environment and incident solar wind to better understand the magnetotail. Study fundamental plasma physics at the fluid-kinetic interface.	High	The behavior of plasmas in the transition from kinetic (particle) to fluid scales is a problem of critical importance to many fields of study. The size of the lunar disk, and of regions of enhanced magnetism on the lunar surface, span the kinetic and fluid ranges of many particle species. This permits a study of fundamental physics at the kinetic/fluid interface to be made.	Requires orbital mission, perhaps as a drop off. Objective-to- architecture rating: [1]
mENVCH7	Characterize the lunar atmosphere to understand its natural state. Of major importance is the electromagnetic and charged dust environment and interaction with the variable space environment.	High	NRC interim report identifies this objective as high priority. Highly likely that electrostatic charging and dust environment will have direct impact on operational mission. Science applications are specifically targeted to the particular nature of the lunar environment and the issues of critical systems and human operations. Safety and reliability designs would require investigation before substantial human activity.	Requires both orbital mission, perhaps as a drop off, and surface lunar package before substantial human activity. Objective-to- architecture rating: [1]
mENVCH10	Map the surface electromagnetic field of the Moon to understand the operational environment of the Moon. Measure the lunar crustal magnetic field and understand its origins and effects.	High	This is a subset of complete mENVCH10 objective. The magnetic field is important for the local plasma, dust, and particle environment. This objective represents new science in unique plasma parameter regimes. It relates to the history of the Moon and an analog for Mars. Magnetic shielding may influence site selection of some exploration activities. Similar instrumentation needed for other objectives.	Orbital in initial stages (low perilune). In-situ rover studies around outpost and during sorties to supplement; selected oriented sample returns. Objective-to- architecture rating: [4]
mENVCH4	Characterize the dust environment at several locations on the lunar surface to better understand the operational environment of the Moon.	High	There is a highly variable plasma environment at the orbit of the Moon due both to the changing conditions of the impinging solar wind and traversals of the magnetosphere. The Moon can enter the hot and tenuous plasma sheet in the Earth's magnetotail, causing increased electrostatic potentials. The resulting surface charging may drive the electrostatic transport of charged lunar dust. The lunar dust-plasma is highly susceptible to space weather. Therefore, we need to observe the dust/plasma environment during range of different solar and magnetospheric activity conditions.	Consider strategic location (South Pole), as well as, or in addition to, distributed sites. Objective-to- architecture rating: [1]

mENV-MON1a	Monitor space weather in real time to determine and mitigate risks to lunar operations. Utilize the coordinated, distributed, simultaneous measurements by the heliospheric great observatory for predictive models of space radiation at the Moon.	High	(1) Mitigating the exposure risk requires the delivery of reliable operational products, based on monitoring of hazardous radiation, to mission operators, planners, and crews. It will also require a dedicated effort to generate near-real-time operational data that are supported by a fundamental understanding of the underlying physics. The infrastructure to monitor space weather over timescales of days - hours - minutes exists. This science is of high intrinsic value because developing such a predictive capability requires the solution of many long-standing problems in heliophysics. High in terms of scientific discovery potential, as well as for practical (operational) considerations. (2) This science objective will probably be achieved only partly by the time of the first lunar landings and will be improved upon continually with more capable instrumentation and higher fidelity models. Nevertheless, the accomplishments will be of high scientific value, and very valuable predictive capabilities will be developed in time to support crewed lunar operations.	Not on the Moon; upstream monitoring measurements must be located on the Sun-Earth line as close to the solar source as is feasible. Objective-to- architecture rating: [5]
mENV-MON1b	Monitor space weather in real time to determine and mitigate risks to lunar operations. Utilize real-time measurements on the Moon to provide redundant forecasting/nowcasting of space weather.	Medium	Although deployment of instrumentation on the Moon for space-weather monitoring is unlikely to yield major scientific advances, even simple full Sun sensors can provide valuable on-site information about the x-ray flux and particle acceleration in the low corona. More detailed imaging instruments can provide a redundant forecasting capability and training for the Mars outpost. These measurements provide direct input to predict the effects on the lunar dust/plasma environment.	Instrument suite can be designed to fit in the existing architecture. A major goal is learning how to run an operational system in a harsh environment. On-site operations need to be carried out by a trained scientist-astronaut at the lunar site, with a view to more independent operation during Mars missions. Objective-to- architecture rating: [5]
mENV-MON2	Monitor lunar environmental variables in real time to determine and mitigate risks to lunar operations. Use real-time observations on the Moon to determine the potential and duration of radiation hazards, the electrodynamic plasma environment, and effects of dust dynamics and adhesion.	High	(1) Monitoring the radiation environment will require dosimetry and a solar proton telescope. It is this telescope that SMD can provide. It must measure protons from 20 to 1000 MeV. In addition to its use for assessing crew radiation exposures, it will provide scientific data for basic research in heliophysics. Further, the Moon's electrodynamic plasma and dust environment must be monitored in real-time to determine electrostatic and dust hazards. (2) The likelihood of successful operation is excellent and the likelihood of achieving science is good. (3) Important for crew safety.	Implementation should be co-located with human operations Objective-to- architecture rating: [1]

mENVCH2	Characterize radiation bombardment at several locations on the lunar surface and subsurface to better understand the operational environment of the Moon.	Medium	(1) The only intrinsic value is the validation of transport code calculations of lunar neutron albedo. It will be helpful to validate the model predictions for the radiation environment on the lunar surface. The biggest uncertainty is thought to be the contribution of neutron albedo to the radiation dose to the crew. Low importance in terms of scientific discovery potential, but important for crew safety. (2) The likelihood of achieving this goal is very high because it relies on the use of well understood and proven detector technologies.	The objective can be completely addressed at a single site. It would have been enough to do it at only one site even if the crew were visiting multiple sites on the Moon. Objective-to- architecture rating: [1]
mGEO9	Understand the nature and history of solar emissions and galactic cosmic rays.	High	The lunar regolith carries a record of the history of solar energetic particles, galactic cosmic rays, and the motion of the heliosphere through the Milky Way. Shaded areas may form cold traps for volatiles. Intrinsic scientific value is high. Samples to be extracted to study lunar geology can be used. However, for dating purposes, samples should be chosen in the context of the lunar stratigraphy. Trenching is the preferred approach. The techniques required for this objective are similar to other lunar regolith survey requirements.	A comprehensive historical picture would require samples illustrating a range of dates, and limitation to a single site may limit the variety of samples available. However, the apparent ubiquity of ejecta layers on the lunar surface indicates a single site should be sufficient. Objective-to- architecture rating: [1]
mHEO1	Image the interaction of the Sun's heliosphere with the interstellar medium to enable identification and comparison of other heliospheres.	Medium	The heliospheric boundary can be imaged from the Moon using energetic neutral atoms, extreme ultraviolet, and soft x-ray fluxes. The study of the global structure of the heliosphere and its interaction with the local interstellar medium is of high value. However, the presence of neutral atoms in the lunar exosphere will cause a significant foreground for Energetic Neutral Atom (ENA) Imaging. Not compelling to do from the Moon.	ENA technique requires remote (satellite) perspective. Other techniques may be implemented on the lunar surface. Objective-to- architecture rating: [1]
mHEO2	Perform low-frequency radio astronomy observations of the Sun to improve our understanding of space weather.	High	Probe particle acceleration in the tenuous upper solar atmosphere and in interplanetary space. This is accomplished by imaging the low-frequency plasma radiation produced by the accelerated particles. An array of small radio telescopes covering spanning tens of km would provide the necessary spatial resolution. Kapton roll deployment technology may revise this assessment upward.	For full sky coverage, multiple sites would be required. Objective-to- architecture rating: [2]
mHEO5	Analyze the composition of the solar wind to improve our understanding of the composition and processes of the Sun. Composition and flux of interplanetary/ interstellar grains should also be considered.	High	(1) Solar wind composition has recently been measured by Genesis, with less than complete success due to its hard return to Earth. Lunar observations would complete the necessary reservoir of samples for 21st century science. (2) The flux and composition of the interplanetary and interstellar grains bombarding the lunar surface are important measurements to both the Heliospheric and the Astrophysical communities, and are a fundamental source of maintaining the lunar atmosphere and modifying the micrometeor-gardened lunar soil.	Observation site requires long intervals of exposure to the solar wind. Objective-to- architecture rating: [1]

mHEO6	Image the interaction of the ionosphere and magnetosphere to understand space weather in the regions of space where most commercial and military space operations occur.	High	Imaging of the geospace environment from the Moon has high intrinsic science value and contributes to operational space weather products. Observations from the Moon give excellent full disk coverage of the Earth unavailable from LEO and GEO orbits. Lunar surface observations of plasma distributions and flow in geospace enable comprehensive diagnostics of space weather processes.	The instrument site must maximize view of Earth. Objective-to- architecture rating: [1]
mHEO7	Perform high-energy and optical observations of the Sun to improve our understanding of the physical processes of the Sun.	High-Energy Observatories - High Optical observatories - Low	Studies of very high energy process require imaging of high energy x ray and gamma rays that cannot be imaged using conventional optics. However, collimators and grid shadowing techniques can provide data that can be used to form images. Grids and detectors must be extremely stable and separated by long distances, which is difficult to achieve in space. The near vacuum and seismically quiet environment of the Moon would allow the construction of an ideal hard x-ray/gamma ray observatory because stability is the primary driver of the design. While scientifically important and essential for safe lunar operations, solar optical observations are better done by a constellation of observatories in Sun-synchronous Earth orbit.	A site a few hundred meters in length in the sunlight would be sufficient. Objective-to- architecture rating: [1]
mHEO8	Analyze the Sun's role in climate change to gain a better overall understanding of climate.	High	The Moon is a platform from which one can measure the three fundamental components of climate (change)—the solar constant, terrestrial reflectance, and Earth's thermal emission. The required technologies are mature and robust. The Moon is not considered to be the best place to measure the solar irradiance, although measurements of the Earth's albedo may be. Measurements of the Earth's albedo fall within the purview of Earth science.	The objective can be completely addressed at a single site. It would have been enough to do it at only one site even if the crew were visiting multiple sites on the Moon.

The realm of heliophysics is the perilous ocean through which explorers, both robotic and human, must journey to reach the dusty shores of the Moon, then Mars.

Authors

Dr. James Spann, NASA Marshall Space Flight Center, Huntsville, AL, *james.spann@nasa.gov*
Dr. Barbara Giles, NASA Headquarters, Washington, DC, *barbara.giles@nasa.gov*

APPENDIX 3: WORKSHOP FINDINGS HELIOPHYSICS SUBCOMMITTEE

APPENDIX 4: WORKSHOP FINDINGS
PLANETARY PROTECTION SUBCOMMITTEE

Context

The PPS of the Science Committee of the NASA Advisory Council is charged with providing advice on planetary protection policy and mission categorization to NASA and the Planetary Protection Officer, in accordance with the Committee on Space Research (COSPAR) guidelines and Article IX of the 1967 Outer Space Treaty.* At the Tempe Workshop, the goal of the PPS was to ensure that planetary protection requirements for preventing biological and organic contamination of solar system bodies will be considered to the greatest extent possible during the development of technologies and procedures to enable human exploration of the solar system, for which a return to the Moon is the first step.

By NASA and COSPAR policy, missions to the Moon are currently considered Category I, which means that operations on the Moon are not constrained by planetary protection restrictions on biological and organic contamination. The Moon is considered to be a sterile and organically clean environment (with potential exceptions such as possible polar deposits of organic materials derived from impactors and sequestered in cold traps), which makes it an optimal location to evaluate the magnitude and range of biological contamination associated with human exploration, as well as to develop technologies designed to mitigate planetary contamination resulting from human presence. A better understanding of organic and biological contamination resulting from past or planned human activities on the Moon will facilitate development and testing of equipment and technologies designed to limit human-associated contamination during exploration of more distant planetary bodies, to which planetary protection restrictions are currently applied.

Considerable experience gleaned from the past four decades of robotic exploration, in addition to early efforts in planetary protection (then called planetary quarantine) during the Apollo program, have demonstrated that planetary protection policies and procedures must be incorporated into mission planning from the very earliest stages. Delaying planetary protection considerations to the later stages of mission design consistently leads to vastly increased costs, damaging schedule delays, and potential loss of missions. Technologies and procedures that will be used during human missions to Mars must be developed and established early in the planning process and tested under realistic field conditions to ensure their compliance with planetary protection policies. By COSPAR guidelines and NASA policy that implement international agreements of the 1967 Outer Space Treaty, missions that do not comply with planetary protection requirements will not be permitted to launch.

Key Findings

During the course of discussions at the workshop, two key issues were raised repeatedly that members of the PPS felt were essential to address during exploration of the Moon in order to prepare for future missions to Mars. A third concern recognized—specific to the Moon—was that exploration of scientifically interesting polar regions on the Moon does increase the possibility of contamination, which in turn might interfere with future scientific discovery. Key findings are listed here:

1. Exploration of the Moon has produced and will produce biological and organic contamination at the sites where human and/or robotic exploration takes place. Operations on the Moon are not constrained by planetary protection restrictions, which makes the Moon an optimal location to establish the magnitude of contamination associated with human exploration and effects of the lunar environment on such contamination over time. Previous lunar exploration efforts, including both robotic missions and the manned missions of the Apollo program, have left behind artifacts on the Moon that are known to contain organic and microbial contaminants. These locations are ideal for testing planetary protection technologies and procedures to detect biological or organic contamination. In addition, the Moon is an excellent testbed for developing and testing technologies for the containment of collected samples, to prevent both forward contamination of the sampling site, and backward contamination of the habitat, return vehicle, and laboratory in which the sample containers will be opened.

2. The Moon is an excellent testbed for developing technologies that may be required to permit human exploration of protected planetary bodies. The lunar return can facilitate development and testing of equipment and technologies designed to limit human-associated contamination. Many processes and technologies required for planetary exploration are likely to produce organic and biological contaminants that are regulated by planetary protection policy. Because organic and biological contamination of the Moon is not restricted, technologies that will be required for exploration of protected locations can be tested and optimized without costly limitations. Necessary technologies that will need optimization to minimize contamination include pressurized habitats and spacesuits as well as robotic and human-associated mobile equipment used for exploration or ISRU. Such technologies and procedures are expected to be required before humans can be permitted to travel to Mars or other protected solar system bodies.

3. Lunar volatiles in polar deposits are susceptible to organic contamination during exploration, and future investigations may indicate that these regions contain materials of interest for scientific research. These regions of the Moon, though currently considered Category I, may be considered for protection at a greater level pending future COSPAR policy discussions.

Detailed Assessment

The spreadsheet provided to participants at the beginning of the Tempe Workshop included only two objectives considered of relevance to planetary protection, mOPS7 (to investigate astrobiology protocols and the search for life), and mOPS8 (to evaluate and improve planetary protection protocols). During discussion of our key findings by the PPS, the two objectives were subdivided to highlight essential components of those activities, and additional topics were also included. Both the old and the revised objectives are listed in the following spreadsheet:

Objective ID Number	Name	Summary	Value	Objective-to-architecture rating
mOPS7	Evaluate astrobiology protocols and measurement technologies that will be used to test for life on other planets.	Evaluate contamination control protocols and establish no-life baselines for scientific technologies that will be used to test for life on other planets.	Astrobiology protocols and technologies can be uniquely tested on the Moon since it is devoid of life. These technologies can be used to test for life elsewhere in the solar system. Operational tests away from the Earth provide more relevant validation of approach.	1
mOPS8	Evaluate planetary protection protocols to develop the next generation planetary protection policy.	Evaluate planetary protection protocols by first characterizing the biological effects of human activity on the lunar surface. Develop and test decontamination of astronauts and equipment returning from the Moon to control forward and backward contamination as a precursor to human return from Mars.	Understanding the impact of human activity on the lunar surface is necessary to develop the next generation of planetary protection protocols. These protocols will help prevent forward environmental contamination of sites on the Moon and backward contamination of crew and cargo returning to Earth. After evaluating these protocols, they can be used as models for future Mars exploration missions.	1
mOPS8.1	Use the Moon and lunar transit/orbits as a testbed for planetary protection procedures and technologies involved with implementing human Mars mission requirements prior to planning these missions.	Evaluate and develop technologies to reduce organic and biological contamination produced by spacesuits, pressurized habitats, and human-robotic interactions. Study lunar spacesuit competency, containment, and leakage issues, and the ability of evolving suit requirements to affect Mars suit, portable life support systems (PLSS), and habitat designs and requirements.	Although contamination of the Moon is not restricted by COSPAR or NASA planetary protection policy, the Moon provides a sterile and organically clean environment in which to evaluate current performance of human exploration technologies, resulting in subsequent improved contamination control as will be required for further planetary exploration in more restricted locations such as Mars.	HIGH
mOPS7.1	Use highly sensitive instruments designed to search for biologically derived organic compounds to perform in situ investigations of lunar landing sites.	Assess the contamination of the Moon by lunar spacecraft and astronauts at a variety of locations.	Valuable "ground truth" data on in situ contamination of samples supports future Mars sample return missions (sample integrity).	HIGH
mOPS7.2	Understand possible contamination of lunar ices by non-organically clean spacecraft.	Evaluate and develop technologies to reduce possible contamination of lunar ices.	Understanding of how spacecraft might contaminate lunar volatiles addresses both mission-science and resource contamination concerns.	HIGH
mOPS8.2	Understand the extent of terrestrial contamination and survival in lunar environments.	Perform chemical and microbiological studies on the effects of terrestrial contamination and microbial survival, both during lunar robotic and human missions.	Dedicated experiments in a variety of lunar environments/depths will facilitate understanding and future remediation of potentially hazardous contamination events. 'Natural' experiments initiated during the Apollo missions could be studied by revisiting the Apollo sites.	MEDIUM

mOPS8.3	Develop technologies for effective containment of samples collected by humans.	Develop technologies for effective containment of samples collected by humans, to feed forward into designs that will help prevent forward and backward contamination during Mars missions.	Technology development for sample collection supports future Mars sample return missions (sample integrity).	MEDIUM
mOPS7.3	Use the lunar surface as a Mars analog site to test proposed life detection systems in a sterile environment that are designed to go to Mars.	This is similar to Viking's Antarctic analog field tests used to ensure a lack of false positives and to evaluate how sensitive the system is to human contamination.	Detection at varying distances from human activity could shed light on movement of materials, which could help establish the distances for "quarantine zones" around special regions.	LOW

Enabling Technologies

A number of discussions took place around the issues of technology required for planetary protection on human missions to Mars and how exploration of the Moon could be useful in the development of such technology. Much of the required technologies have been or are being developed for the robotic space program or as commercial products; however, additional work will be required to adapt available products for use during human spaceflight missions. In addition, a number of technologies required for long-duration human life support are not yet mature and will be quite costly to develop further. Considerable effort should be expended to ensure that life support and habitat technologies developed for the Moon are usable for later human missions to other solar system bodies that have more stringent planetary protection requirements. Details on three specific topics of discussion are provided below.

1. A substantial amount of technology relevant to planetary protection and other scientific questions has been developed by NASA through the advanced technology programs—Astrobiology Science and Technology for Exploring Planets (ASTEP), Astrobiology Science And Technology Instrument Development (ASTID), Mars Instrument Development Program (MIDP), Planetary Instrument Definition and Development Program (PIDDP), etc. Presentations given by PPS-invited speakers described several instruments developed for robotic spacecraft exploration that have been adapted to interface with humans, either with the assistance of a robot or through direct operation while wearing a spacesuit. Such instruments have been operated successfully in remote locations on Earth, such as Svalbard Island and Antarctica. These technologies and instruments, which include robotic sample collection and sensitive, rapid assay methods using field-portable equipment, should be reinvestigated for relevance to human exploration requirements.

2. However, commercial off-the-shelf technologies are not rated for spaceflight, and the modifications necessary would require expensive retooling. For example, the Lab-on-a-Chip Application Development-Portable Test System (LOCAD-PTS) instrument that is currently being flown on the ISS required complete reengineering to accommodate man-rated space flight requirements, such as low outgassing from construction materials, radiation-resistant electronics, etc. De novo

development of necessary technologies required for long-duration human space flight missions is likely to prove at least as cost-effective as modification of existing commercially available equipment.

3. The Moon should be used as a testbed of advanced life support systems for Mars exploration. There should be a move towards sustainable high efficiency closed-loop systems, as well as a comprehensive effort to qualitatively and quantitatively assess their effectiveness.

Issues

The PPS has identified several issues that would benefit from additional attention during planning of the Constellation Architecture. The near-term focus on exploration of the Moon affords a unique opportunity for testing planetary protection protocols in a challenging space environment known to be sterile but not restricted by planetary protection policy. Every effort should be made to take advantage of this opportunity to ensure that planetary protection protocols are established to the extent that will be required for future human missions to solar system bodies receiving more than Category I protection.

A separate, follow-on meeting to explore opportunities in biological sciences in partial gravity and at a pressurized lunar outpost is suggested. Such a meeting will continue and expand the effort started two years ago that brought together planetary protection experts, astrobiologists, life support specialists, and engineers to discuss human exploration of space. Additional meetings should address, in a systematic and detailed fashion, cross-cutting science and technologies that are both enabled by the lunar exploration program and will enable human exploration to more remote solar system bodies.

Substantial proportions of the lunar dust are submicron-sized and could pose a significant health hazard, although no adverse effects have been detected due to the limited dust exposures of the Apollo astronauts. Current efforts to use data from Apollo and terrestrial dust exposure studies should be strongly encouraged to better understand exposure times, particle distributions, particle morphology, chemistry and reactivity that may a pose a problem. Human health must be assessed routinely during exposure to planetary environments to evaluate the potential risks upon return to Earth.

A variety of equipment is available or under development that would be desirable to test on the Moon for studies relevant to human health and planetary protection, and field-capable versions will certainly be completed prior to the first human return to the Moon. In planning the lunar outpost, it will be very important to include sufficient allotments for up-mass to the lunar science laboratory that facilitate testing of planetary protection technologies and experimental equipment. In addition, outpost crews will need appropriate training in operation of the equipment, and sufficient time scheduled to allow the necessary testing and experiments to be performed.

Planetary protection technologies to reduce contamination from human missions must be supported at an appropriate budget level if human missions to Mars are to be properly planned and implemented.

Effective communication with the public about planetary protection goals and requirements will be important to garner public support for both robotic and human missions to other planetary bodies.

Authors

Catharine A. Conley, NASA Science Mission Directorate, *cassie.conley@nasa.gov*
Nancy A. Budden, Office of the Secretary of Defense, *nancy.budden@osd.mil*
John D. Rummel, NASA Headquarters, *john.d.rummel@nasa.gov*

---

* **Article IX of the 1967 Outer Space Treaty and COSPAR guidelines for Planetary Protection**

Article IX of the Outer Space Treaty of 1967,** to which the United States is a party, states in part that
"...parties to the Treaty shall pursue studies of outer space, including the Moon and other celestial bodies, and conduct exploration of them so as to avoid their harmful contamination and also adverse changes in the environment of the Earth resulting from the introduction of extraterrestrial matter and, where necessary, shall adopt appropriate measures for this purpose..."
These basic treaty principles are not elsewhere defined in the treaty itself, but like other treaties this treaty is "the supreme law of the land" under the U.S. Constitution (Article VI).

To ensure treaty compliance, and upon the repeated recommendations of the U.S. National Academy of Sciences, NASA maintains a planetary protection policy to protect against biological or organic contamination that might jeopardize scientific exploration or the safety of the Earth's environment. NASA also works with COSPAR, an interdisciplinary committee of the International Council for Science that consults with the United Nations in this area, to ensure that there is an international consensus policy that can be used as the basis for planetary protection measures to be taken on international cooperative missions. In general, NASA will approve the flight of NASA-developed instruments and/or experiments on non-U.S. planetary spacecraft only if the launching organization adheres to the COSPAR-approved planetary protection policy and its requirements (as noted in NPR 8020.12C).

**("Treaty on Principles Governing the Activities of States in the Exploration and Use of Outer Space, Including the Moon and Other Celestial Bodies" entered into force October 10, 1967. 18 U.S. Treaties and Other International Agreements at 2410-2498.)

APPENDIX 4: WORKSHOP FINDINGS PLANETARY PROTECTION SUBCOMMITTEE

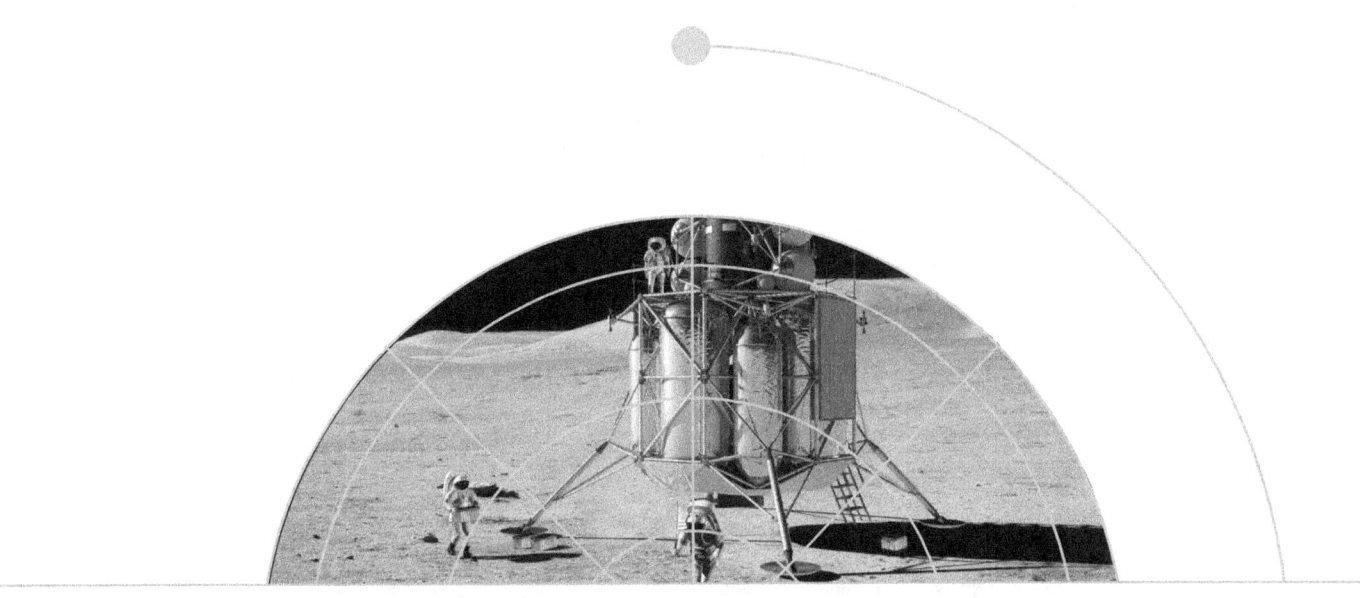

APPENDIX 5: WORKSHOP FINDINGS
PLANETARY SCIENCE SUBCOMMITTEE

Presentations to the Planetary Sciences Subcommittee (PSS) breakout sessions at the lunar architecture workshop focused on the major science themes developed by the lunar architecture team and modified by the LEAG. The interim report of the Space Studies Board on lunar science priorities also was considered. Discussion on Thursday afternoon included input from the PSS and input from the general audience, which fluctuated between about 65-90 participants (depending on the topic under discussion). Discussion focused on the objectives and how they would be achieved within the current lunar architecture, noting what modifications would be needed and what technology developments will need to be focused upon.

Objectives Grouped under Five Overarching Themes

The Planetary Sciences Subcommittee breakout sessions at the workshop examined all 16 of the GEO-SAT objectives and grouped them under five broad science themes. These are indicated below, along with the objectives that are grouped under each one.

- Investigation of the geological evolution of the Moon and other terrestrial bodies (mGEO-1, mGEO-2, mGEO-3, mGEO-5)

- Improved knowledge of impact processes and impact history of the inner solar system (mGEO-6, mGEO-7, mGEO-8)

- Characterization of regolith and mechanisms of regolith formation and evolution (mGEO-9, mGEO-10, mGEO-11, mGEO-14)

- Study of endogenous and exogenous volatiles on the Moon and other planetary bodies (mGEO-4, mGEO-12, mGEO-13)

- Development and implementation of sample documentation and return technologies and protocols (mGEO-15, mGEO-16)

On the basis of the overarching themes and subsidiary objectives, the LEAG is charged to correlate the objectives to an implementation plan. Correlation will include measurement objectives, geographic coverage, and sampling and documentation strategies. Objectives will be distinguished on the basis of major progress that can be made through the current exploration architecture.

Discussion of each of the LEAG GEO-SAT (Geoscience Objectives Special Action Team) Objectives as Modified by the LAT

Introduction:

Science themes that were assembled by the LAT from the ESMD Lunar Exploration Workshop (April 2006) were ranked by the LEAG as the GEO-SAT for lunar-science relevance and by the Mars Exploration Program Analysis Group (MEPAG) for relevance to the exploration of Mars. During the Tempe workshop, the priorities of these themes were debated and ranked. Implementation of the science was discussed in detail, leading to the recommendations to NASA. Here we list, in order of the GEO-SAT/LAT objectives, the science themes, their ranking, and a summary of the discussion. A table summarizing the objective assessments and rankings follows this section. Note that slight adjustments to the titles of mGEO-6, -7, -10, -12, -13, and -15 in the mGEO-SAT document have been made to clarify the science objectives.

mGEO-1: Determine the internal structure and dynamics of the Moon to constrain the origin, composition, and structure of the Moon and other planetary bodies.

PSS/LEAG Score for Lunar Science Objectives (1 = Low; 10 = High): 10
MEPAG Ranking (feed-forward to Mars Science objectives): High

Discussion:

- This objective has received high rankings from the LEAG, MEPAG and LAT.
- Technology development is needed to create a common geophysics package that can be deployed robotically or by astronauts during any mission to the lunar surface. Technology to deploy such instrumentation from orbit is also needed.
- A long-lived (>6 years), low-mass power supply is needed. If the network is built up incrementally, the initial stations still need to have long life spans.
- Different numbers and placements of seismic stations are needed to accomplish different objectives. Examples are given below:
    o Two Stations: This is the minimum number, deployed antipodal to each other with one being close to a known, reliable seismic source (e.g., close to the A-1 deep moonquake nest or the far-side A-33 nest). A network of two seismometers will yield only approximate information on the locations of deep moonquakes, and little to no information on the origin of shallow moonquakes or crust/mantle heterogeneity.
    o Three Stations: the minimum number of stations to locate and time each deep moonquake, but these need to be dispersed over a much wider area than those deployed during Apollo (including a station on the far side). Data from three stations will be sufficient to determine approximate meteoroid impact times and locations. With smaller station spacing, smaller impacts can be detected by all three stations, whereas with larger station spacing, a larger area can be covered for detection. As with a two-seismometer network, a network of three seismometers will yield

only approximate information at best on the locations of shallow moonquakes, and little information on crust/mantle heterogeneity.
- o Four Stations: Exploring lateral heterogeneity in the lunar crust and mantle requires a minimum of four seismometers, but this depends on the distribution of the stations (to obtain global distribution of structural and seismic-velocity variations, a globally distributed array of seismic stations is required). There is no clear limit to the number of seismic stations needed to do this, but the larger the number of stations, the more detailed the result will be.
- o > Four Stations: A larger number of seismometers is required to determine source depths for shallow moonquakes because a smaller spacing of stations is needed relative to that required for deep sources. What is required is to place clusters of seismometers at a number of the approximated shallow moonquake locations, as well as any proposed lunar habitat site. Such a cluster could be set up at one site using the current architecture, then at one location (at least), a cluster of three seismometers could be set up to answer some of the fine-scale questions raised above.
- o General: Any network should have a broader coverage than that of the Apollo Passive Seismic Experiment (PSE) network. Many of the shortcomings of the Apollo seismic database stem from the lack of station coverage beyond the near-side center of the Moon. Thus, extending the station coverage should be the primary objective of our next lunar seismic observation. Also important is the length of observation; that is, longer duration experiments are needed, preferably longer than the 5–8 years that the Apollo stations were operational.
- A bare-bones global network (e.g., four stations with much wider coverage than Apollo) set up prior to the seventh human landing would add greatly to meeting this objective and assist in the proper planning for activities related to that landing (see above).
- Orbital geophysical data are also important, such as gravity, magnetics, and the composition and dynamics of the lunar atmosphere.
- A range of geophysical properties need to be measured over a number of years in order to achieve this objective (e.g., heat flow, magnetism, seismic events and their magnitudes, locations and travel times, and dynamics of responses).
- This objective cannot be addressed from a single site (if mobility is limited to 20 km). However, a geophysical station (seismometer, heat-flow probe[s], magnetometer, for example) should be set up at an outpost site because it would provide the following: a record of seismicity at the outpost site; some limited information about the interior; and, most importantly, it would represent the initial node of a long-duration, global seismic network.
- To ultimately fulfill this science objective, access to sites across the entire Moon is essential. If global access is available within the outpost architecture, this objective will be achievable.
- NASA is encouraged to consult with international partners to ensure that any mission to the Moon's surface deploys a common geophysical package.

mGEO-2: Determine the composition and evolution of the lunar crust and mantle to constrain the origin and evolution of the Moon and other planetary bodies.

PSS/LEAG Score: 10
MEPAG Ranking: Medium

Discussion:
- Overlaps with mGEO1, but requires targeted sample returns from impact basins (especially the South Pole-Aitken Basin), vent crater deposits, pyroclastic deposits, central peaks of impact craters, ancient buried lava flows or "cryptomare," volcanic "red spots" (areas likely to represent compositionally evolved volcanic materials), impact crater ring exposures, far-side and polar crust and mare basalts, and unsampled near-side mare basalts (including the "youngest" basalts). As with mGEO-1, global access from an outpost could enable this objective to be achieved.
- It is unlikely that the current architecture will allow this objective to be achieved in its entirety in the near term, although important insights may be gained by sampling bedrock materials and ejecta present in the vicinity of an outpost site. However, depending on the geological setting of the specific site chosen, significant progress could be made by intensive study of the rock components present in the regolith and in crater ejecta at one site. This could be accomplished by returning significant amounts of regolith through the region surrounding the outpost (see also mGEO-5), and perhaps by using automated techniques to screen samples on the surface. Proximity to a diversity of geologic terrains is particularly important.
- Robotic missions with/without humans present will play an important part in achieving the scientific goals—especially in sampling relatively small, issue-critical sites identified from orbital data—and in allowing full "global" access. Robotic missions also will play a role in deploying global network instruments. However, robotic sampling does not satisfy all sampling needs for documentation of geological context—sample context will be neglected unless robots work in unison with humans (telepresence) to analyze or sample variable/large terrains. Even with human remote telepresence, subtle details of sample context and much of the spontaneity of follow-up observations that humans on the spot would provide will be lost.
- Technology development: develop sample return mechanisms for robotic (simple) and human (complex) sampling sorties. There are some sites that can be sampled robotically and those that need human presence (or the combination of humans and robots).
- Development of technology for efficient human exploration (mobility and pressure suit systems), observation (mapping, active geophysical and geochemical sensors and geometric and geotechnical measuring systems), and sampling and sample documentation of complex sites.

mGEO-3: Characterize the lunar geophysical state variables to constrain the origin, composition, and structure of the Moon and other planetary bodies.

PSS/LEAG Score: 9
MEPAG Ranking: Medium

Discussion:
- Geophysical state variables include the lunar gravitational potential field, heat flow, lunar rotational fluctuations, lunar tides and deformation, and the present and historic magnetic fields. Geodetic information about the Moon can be used to determine global-scale geophysical characteristics that include the core and deep mantle.
- Long distance surface mobility and global access will enable this objective to be achieved. An analytical capability to determine geophysical/geotechnical parameters that would be useful for outpost development is desirable.
- This objective is also enabled by a combination of sample return (or collection and characterization of surface physical samples) for ground truth and orbital measurements. Orbital measurements (e.g., magnetic measurements of Moon in free solar wind, magnetosheath, and magnetotail) are important for electromagnetic sounding and conductivity measurements to constrain the size and nature of the lunar core.
- Knowledge of the heat flow of the Moon in a global sense is needed for proper interpretation of other geophysical data. Key areas for heat-flow measurements include the hot spot on the western near side in the Imbrium-Procellarum region, the interior of the far-side South Pole-Aitken basin, and a location within the feldspathic highlands away from the regions of high thorium concentration.

mGEO-4: Determine the origin and distribution of endogenous lunar volatiles as one input to understanding the origin, composition, and structure of the Moon and other planetary bodies.

PSS/LEAG Score: 7
MEPAG Ranking: Low

Discussion:
- Endogenous volatile deposits are not present everywhere, so surface mobility from an outpost will be an enabling capability for this objective.
- Some high priority aspects of this objective will require sample return, such as from pyroclastic deposits/cryptomare. Field-work and surface-mobility capabilities with local (~50 km), regional (up to 500 km), and global access will enable this objective to be achieved. Robotic sample return is possible. Without this capability, the objective is unlikely to be achieved because there are not substantial pyroclastic deposits known to be at the poles.
- Fieldwork will allow a better understanding of the current outgassing environment through visits to sites that are thought to be areas of active or geologically recent outgassing.
- Endogenous and exogenous volatiles need to be differentiated. Exogenous volatiles will include trapped and implanted components.

mGEO-5: Characterize the crustal geology of the Moon via the regolith to identify the range of geological materials present.

PSS/LEAG Score: 9
MEPAG Ranking: Low

Discussion:
- This is more than simply a regolith study. It requires rocks from the regolith to be sampled and returned from a variety of locales, including the far side, because a variety of samples will be needed to examine the diversity of the lunar crustal rocks. Sample return could be accomplished by human missions, robotic missions with significant mobility, or a combination of both. However, this objective could be initiated by a single sample of regolith (to look at the diversity of ejecta material in it) that can be obtained from anywhere, including an outpost location.
- The discussion centered on this objective being intimately linked with mGEO-2.
- Integration with orbital geochemical and geophysical data is vital to achieve this objective.
- Assisting in meeting this objective means routinely collecting "sortie rake samples" and "contingency" samples at various locations during the exploration of or landings in new regions.
- Extensive fieldwork enabled by global surface mobility, along with sample characterization and documentation capabilities in the field and at an outpost, are enabling for this objective.
- Robotic sortie missions could fully meet the objective in locales where human missions are unlikely to land.

mGEO-6: Characterize the impact process, especially for large basins, on the Moon and other planetary bodies to understand this complex process.

PSS/LEAG Score: 9
MEPAG Ranking: Low

Discussion:
- mGEO6 is process-oriented, but relates directly to processes active and important on the early Earth and throughout the planet's history.
- A lunar outpost is a good place to begin addressing this objective, particularly if located on a ring of the South Pole-Aitken basin.
- Shallow geophysical studies will allow investigations of the three-dimensional structure of craters and should be part of any study of impact processes.
- Significant progress can be made at a single site by studying a number of impact craters in detail; however, local to regional surface mobility for astronauts would be needed.
- Achieving this objective requires orbital and sample data, including geological and geophysical field studies, and the return of key samples to Earth.

mGEO-7: Characterize impact flux over the Moon's geologic history to understand early solar system history.

PSS/LEAG Score: 10
MEPAG Ranking: High

Discussion:
- mGEO7 is history-oriented, relating directly to the history of the Earth and the origin and evolution of life on Earth.
- Originally the objective read "Characterize impact cratering over the Moon . . ."
- Sample return of impact melt rocks from various craters will be needed for precise age dating.
- If the outpost is located within a large basin not previously sampled, significant progress could be made. For example, South Pole-Aitken basin is a very good place to start, but would require a far-side, southern hemisphere site. A location within the South Pole-Aitken basin would provide the access needed to sample its melt sheet, which could date the event as long as the melt sheet can be identified, as well as those of superposed younger basins.
- Surface mobility is an enabling technology in order to gain access to and samples from the largest impact basins.
- Impact-melt samples will need to be returned to Earth for age dating.

mGEO-8: Investigate meteorite impacts on the Moon to understand early Earth history and origin of life.

PSS/LEAG Score: 7
MEPAG Ranking: Low

Discussion:
- mGEO8 resulted from the combination of three different, but related, topics: (1) determine timing and composition of impactors to study the impact history of the Moon, (2) look at the cratering flux and regolith in specific lunar craters, and (3) search for material/impact ejecta from Earth and other bodies to research characteristics of the earlier impact history (i.e., terrestrial meteorites).
- This is an important lunar science objective that is enabled by extensive surface mobility and field-work. The low PSS/LEAG score reflects the lack of confidence in finding early Earth meteorites on the Moon.

mGEO-9: Study the lunar regolith to understand the nature and history of solar emissions, galactic cosmic rays, and the local interstellar medium.

PSS/LEAG Score: 9
MEPAG Ranking: High

Discussion:
- Meeting this objective will require drilling and/or trenching of the lunar regolith, field observations, and sample return, recognizing that agitation of collected regolith will release significant amounts of solar-wind-implanted volatiles.
- This objective would be best achieved if a site can be found where fossil regolith occurs between lava flows or definable ejecta blankets so the enclosing layers can be dated and thus the age of the regolith can be constrained. Such stratigraphy may be best preserved in special environments (especially volcanic terrains) that may or may not be present near an outpost location. Extensive surface mobility would, therefore, be an enabling capability for achieving this science objective. Although the requirement for mobility is not essential to achieve some aspects of this objective, it would certainly enhance the science return.
- This objective should include a study of the megaregolith, and shallow geophysical studies would be useful in defining megaregolith thickness at least at the local scale.

mGEO-10: Determine lunar regolith properties to understand the surface geology and environment of the Moon and other airless bodies.

PSS/LEAG Score: 7
MEPAG Ranking: Low

Discussion:
- This objective refers to regolith properties anywhere, including cold traps.
- These include geochemical, petrologic, and geotechnical properties. The latter will be important in understanding the transmission of seismic energy as wells and the engineering and economic aspects of construction and resource extraction.
- There was discussion of the use of a local active seismic network, including seismic tomography, near the outpost to aid in the understanding of regolith properties in this area, along with ground-penetrating radar. This type of infrastructure would be very useful in achieving the goals of this scientific objective, and this is enabled by an outpost architecture.

mGEO-11: Characterize the lunar regolith to understand the space weathering process in different crustal environments.

PSS/LEAG Score: 7
MEPAG Ranking: Low

Discussion:
- In order to understand the space weathering process over time, regolith of different ages needs to be identified and sampled.
- The phrase "in different crustal environments" was added at the end of this objective title through consensus after discussion regarding the specific goals that this objective should encompass.

- There are two ways this objective can be achieved: (1) trenching and detailed sampling of different levels, and (2) identifying the spectrum of features at different ages (i.e., fresh features vs. degraded features, which could result in the need to sample from widely spaced sites on the Moon, although small craters of a range of ages are present at all locations). The latter would not require trenching but would be enabled by the availability of surface mobility. Trenching and detailed sampling of different levels within the regolith could be suited to the outpost architecture if the maturity (degree of exposure to space weathering) of the regolith changes with depth.
- Detailed field observations (i.e., in-the-field, in situ analyses by means of detailed observations and hand-held instruments) will be needed.

mGEO-12: Characterize lunar volatiles and their sources to determine their origin and to reveal the nature of impactors on the Moon.

PSS/LEAG Score: 8
MEPAG Ranking: Medium

Discussion:
- The words "and their sources" were added to the title of this objective.
- This objective is aimed at understanding cold-trap volatiles (cometary, solar wind implanted, etc.)—how they were deposited, what was their source, and how they accumulated at the poles.
- In situ cold-trap analyses may be required to fully achieve the goals of this objective. Robotic technology developments for operation in extremely low temperatures are needed.
- Sample return may not preserve the integrity of the cold-trap samples unless specialized sampling and containment techniques are developed.
- Although this objective is extremely important scientifically, significant enabling technology developments will be needed to ensure this objective can be successfully achieved.

mGEO-13: Characterize transport of lunar volatiles to understand the processes of polar volatile deposit origin and evolution.

PSS/LEAG Score: 7
MEPAG Ranking: Low

Discussion:
- Goals of this objective could be achieved with orbital spectrometers or a network of surface spectrometers to monitor the exosphere at various places on the lunar surface.
- The timing of this is critical for recording the state of the lunar exosphere and volatile transport process before lunar missions start landing regularly and disturbing the exosphere.
- Transport of volatiles may be related to the source as well as many other variables, such as latitude, magnetospheric phase, mineral composition of the regolith, pick-up ion migration, etc.

- Volatiles, in part, will migrate from cold spot to cold spot rather than migrating off surface, although some will be entrained as solar wind pick-up ions and either lost to space or re-implanted in the regolith. Workshop participants agreed that such transport processes are important but not well understood. Orbital mass spectrometers would be helpful, but no substitute for ground truth from a surface spectrometer network. This could be integrated with a geophysical network deployment.
- The success of this objective is enabled by having global access from an outpost location.

mGEO-14: Characterize volatiles and other materials to understand their potential for lunar resource utilization.

PSS/LEAG Score: 7
MEPAG Ranking: Low

Discussion:
- The previous title, "Characterize potential resources to understand their potential for lunar resource utilization" was changed to the title shown above.
- Any precursor missions are potentially important for carrying ISRU demonstrations, but are likely not essential to achieve the goals of this objective.
- There was no agreement on whether near-surface geophysics (e.g., ground-penetrating radar) or drilling and/or trenching with in situ analysis would be the best way to characterize regolith resources.
- While sample return would yield important scientific discoveries, in situ analyses to characterize resource potential need to be considered and developed.
- NASA and the scientific community must make the best use of orbital data sets (e.g., Clementine, Lunar Prospector, LRO, Chandrayaan [Indian mission], SELENE [Japanese mission], and Chang'e-1 [Chinese mission]) to identify the locations of the sites that have the most potential for resources, as well as detectable and quantifiable surrogates (indicators) to determine the detailed distribution of resources.
- The success of this objective is enabled by having global access from an outpost location.

mGEO-15: Provide curatorial facilities and technologies to ensure contamination and environmental control for lunar samples.

PSS/LEAG Score: 10
MEPAG Ranking: Low

Discussion:
- NASA needs to take advantage of the Apollo experience and knowledge base, especially at the astromaterials sample curatorial facilities at Johnson Space Center.
- This objective is related to science decisions on what kinds of sampling techniques and sample curation/protection need to be done before samples are returned in order to prevent changes in properties during return trips or when opened on Earth.

- When bedrock samples are taken, mechanisms are needed to preserve knowledge of the orientation of the samples, and environmental variables for environmentally sensitive samples must be fully documented.

mGEO-16: Provide sample analysis instruments and protocols on the Moon to analyze lunar samples before returning them to Earth.

PSS/LEAG Score: 9
MEPAG Ranking: Medium

Discussion:
- Consensus was to add "and protocols" between "instruments" and "on the Moon" in order that technologies and instrumentation that facilitate better field practices (e.g., identification, sampling, documentation, and traverse planning) would be considered.
- Analyses of samples before they are returned would allow high-grading of samples to be returned, because the mass that can be brought back under the current architecture appears to be limited. However, this should not imply that all samples collected need to be analyzed prior to return to Earth, and the difficulties in making judgments or analyses that can only be done on Earth should not be underestimated.
- Any analysis at the outpost requires careful protocol development to ensure the integrity of the samples.
- Protocols need to be developed (or refined) for sample collection.
- This has been deferred to CAPTEM for detailed study.

Table 1: PSS Objectives Summary (Assessments and Rankings)

Objective number	Objective description	LEAG/ PSS ranking (1-10) 10: highest priority	MEPAG low-high feed fwd to Mars	Implementation considerations	Rating for polar outpost	Comments
mGEO-1	Determine the internal structure and dynamics of the Moon to constrain the origin, composition, and structure of the Moon and other planetary bodies.	10	high	long-lived power supply; multiple sites widely separated; potential international component	4	This objective cannot be addressed from a single site. However, a seismic station (geophysical station) should be set up at an outpost site because it would provide some information about the interior and, most importantly, it would represent a start toward establishing a long-duration global seismic/geophysical network.

mGEO-2	Determine the composition and evolution of the lunar crust and mantle to constrain the origin and evolution of the Moon and other planetary bodies.	10	medium	targeted sample returns; multiple locations	3	Significant progress can be made by intensive study of one site and documentation and return of rock and regolith samples throughout the region surrounding the outpost. How much progress can be made depends on the geological setting of the specific site chosen; proximity to a diversity of geologic terrains is particularly important.
mGEO-3	Characterize the lunar geophysical state variables to constrain the origin, composition, and structure of the Moon and other planetary bodies.	9	medium	long-range surface mobility; multiple locations; sample return; coordinated remote sensing	4	Little progress can be made on this objective from a single site, with the exception of a heat flow measurement. The utility of a single heat-flow measurement depends on the geological and geophysical setting of the site.
mGEO-4	Determine the origin and distribution of endogenous lunar volatiles as one input to understanding the origin, composition, and structure of the Moon and other planetary bodies.	7	low	ong-range surface mobility; targeted sample returns; volcanic site	4	Achieving this objective requires landing sites with the best chance of yielding significant information about lunar endogenous volatiles, such as pyroclastic deposits, near volcanic vents, or sources of possible recent outgassing.
mGEO-5	Characterize the crustal geology of the Moon via the regolith to identify the range of geological materials present.	9	low	multiple, widely separated sample locations	2	This is less effective than going to diverse terrains on the Moon to sample the crust, but significant progress can be made at one site. South polar location is a previously unsampled terrain. Regolith samples and rock fragments in the regolith complement any collection of large rock samples. Regolith sampling can be done robotically.
mGEO-6	Characterize the impact process, especially for large basins, on the Moon and other planetary bodies to understand this complex process.	8	high	local to regional surface mobility for astronauts; sample return	2	Significant progress can be made at a single site by studying one or more craters in detail. Requires orbital and sample data, and geological and geophysical field studies.
mGEO-7	Characterize impact flux over the Moon's geologic history, to understand early solar system history.	10	high	sample return for age dating; long-range surface mobility and/or access to multiple locations	3	If the outpost were within a large basin not previously sampled, significant progress could be made. For example, if the site were inside South Pole-Aitken basin, it would be possible to sample its melt sheet (hence be able to date the event) and those of superimposed younger basins. Access to South Pole-Aitken basin requires a far-side, southern hemisphere site.

mGEO-8	Investigate meteorite impacts on the Moon to understand early Earth history and origin of life.	7	low	surface mobility; extensive site field geologic investigation; sample return for dating & geochemistry	2	Requires access to multiple impact craters and regolith samples. Well addressed at a single outpost site where numerous craters can be explored and large amounts of regolith can be processed and techniques employed to search for key indicator minerals or chemical compositions.
mGEO-9	Study the lunar regolith to understand the nature and history of solar emissions, galactic cosmic rays, and the local interstellar medium.	9	high	drilling/trenching of the lunar regolith; best done where interlayered volcanics provide age record	3	Extensive regolith excavation at a single site could address this objective by identifying layers deposited by specific impact events. Extensive ISRU processing could aid this search.
mGEO-10	Determine lunar regolith properties to understand the surface geology and environment of the Moon and other airless bodies.	7	low	extensive study of regolith, including excavation, sampling, & geophysical studies	1	This objective can be achieved well at an outpost site. Investigation would go far beyond what is known from Apollo cores and active seismic measurements, and could involve in situ measurements of many geotechnical and other regolith properties. Enabling for exploration.
mGEO-11	Characterize the lunar regolith to understand the space weathering process in different crustal environments.	7	low	local surface mobility; trenching; sample documentation, collection, and return to Earth	1	Can be done well at a single site with detailed investigation of regolith at different locations and with different degrees of surface exposure.
mGEO-12	Characterize lunar volatiles and their source to determine their origin and to reveal the nature of impactors on the Moon.	8	medium	in situ analysis of volatile deposits; operation in extremely low temperatures	1	Analysis of volatiles in the lunar exosphere and in and near polar cold traps are well enabled by a polar outpost location. Needs to be done early in the human exploration program.
mGEO-13	Characterize transport of lunar volatiles to understand the processes of polar volatile deposit origin and evolution.	7	low	global access (range of latitudes & locations) desired	2	Much of this objective can be achieved at a polar outpost site through access to permanently shaded craters and regolith near to and at a range of distances from the pole.

mGEO-14	Characterize volatiles and other materials to understand their potential for lunar resource utilization.	7	low	linked to ISRU; exploration enabling; needs to be phased early; access to specific sites widely separated around Moon	4	Ground truth/in situ characterization of deposits located from orbital data can lead to accurately targeted locations on the Moon. Should be done during the robotic precursor phase to identify the best outpost location. Doing this from a polar outpost location instead of during the precursor phase will characterize the deposits at the site, but this is too late to influence optimal outpost location, thus ranked a "4."
mGEO-15	Provide curatorial facilities and technologies to ensure contamination control for lunar samples.	10	low	development of sample documentation, collection, environmental and orientation controls needed	1	Objective can be well achieved at an outpost location; potential polar volatile deposits provide test case for extremely environmentally sensitive sample documentation, collection, transfer, and processing.
mGEO-16	Provide sample analysis instruments and protocols on the Moon to analyze lunar samples before returning them to Earth.	9	medium	none	1	Objective can be well achieved at an outpost location.

Recommendations

Geophysical Networks

Achievement of several of the highest-ranked lunar science objectives requires the deployment of long-lived geophysical monitoring networks. Precursory technology investments are needed, e.g., development of a long-lived power source and a deployment strategy for stations that are part of such networks. Networks could be built up in partnership with other space agencies provided that a framework for compatible timing and data standards is established. The tradeoff between station lifetime and the timeframe for network deployment should be fully explored.

Background Information: Geophysical networks (i.e., networks of packages containing, for example, a seismometer, heat-flow probe[s], and a magnetometer) are highly ranked in the interim NRC report and from community input as provided by the LEAG. Such networks need not be limited to geophysical instruments, but also include mass spectrometers for exosphere monitoring. However, such networks need to be long-lived (>6 years, which encompasses one lunar tidal cycle) and requires the development of a power source that can achieve this and survive the lunar night.

Sample Return

Achievement of several of the highest-ranked scientific objectives requires the development of a strategy to maximize the mass and diversity of returned lunar samples. The PSS views the 100 kilograms total return payload mass allocation (including containers) in the current exploration architecture for geological sample return as far too low to support the top science objectives. The PSS requests that CAPTEM be asked to undertake a study of this issue with specific recommendations for sample-return specifications to be made by May 1, 2007. The PSS recommends that NASA establish a well-defined protocol for the collection, documentation, return, and curation of lunar samples of various types and purpose in order to maximize scientific return while protecting the integrity of the lunar samples.

Background Information: Collection and return to Earth of lunar samples is vital for science. Sample return has been achieved by the Apollo and Luna missions. The protocols that worked during that time should be enhanced, whereas those that did not need to be revisited so that the lessons learned from Apollo are incorporated into an overall sample strategy. Finally, integration of new field exploration technology will need to be incorporated into this strategy. Technology development in terms of vacuum seals, drive tube extraction, and remote robotic sample return (i.e., direct to Earth without involving the outpost) is a necessity for a number of types of sample investigation. The input from CAPTEM regarding the return sample mass allocation will be important for achieving the science objectives described above.

Astronaut Training

As part of the developing lunar exploration architecture, the PSS recommends that extensive geological, geochemical, and geophysical field training be established as an essential component in the preparation of astronaut crews and the associated support community for future missions to the Moon. Training should involve experts and experience from the non-NASA community, as well as NASA personnel of significant background and experience in field exploration and space mission planning and execution. The training program developed for the Apollo 13–17 missions should be considered a starting point for training of the next generation of lunar explorers. Crews for future lunar missions should include astronauts with professional field exploration experience. Research and training or operational simulations are needed to determine how robots can best be used to assist humans in activities associated with the lunar exploration architecture, which feeds forward to human exploration of Mars.

Mobility

To maximize scientific return within the current exploration architecture, options should be defined and developed for local (~50 kilometers), regional (up to 500 kilometers), and global access from an outpost location. It is important that access to scientifically high-priority sites not be compromised by mobility limitations, both for outpost and sortie missions.

Background Information: The outpost architecture will allow the goals of many more of the science objectives to be achieved as long as sites other than those in the immediate vicinity (1-2 kilometers) of the habitat are accessible. Options for local (~50 kilometers), regional (up to 500 kilometers), and global access from an outpost location should be explored and presented to the Council for review.

Robotic Missions

Robotic missions are highly desirable to carry out many of the highest-priority lunar science objectives. In addition, workshop participants agreed that robotic precursor missions beyond LRO are important for both basic and exploration science (e.g., determining seismicity in proposed outpost locations and defining the nature of the cold-trap volatile deposits). To fully achieve the highest-ranked lunar-science objectives, continued robotic sortie missions will be needed both before and after human presence is established.

Background Information: The overall lack of science associated with lunar missions for a 10-year period of time severely affects future generations of planetary scientists in the following ways. First, "passing the baton" between the Apollo generation and the new generation will not occur in the manner it should—lessons learned from Apollo will have been forgotten. Second, the outstanding young planetary scientists will not have the detailed knowledge of lunar science that has been establish by and since Apollo when we finally return to the Moon near the end of the next decade.

CEV SIM Bay

The PSS recommends that the CEV have a capability similar to the Apollo SIM to facilitate scientific measurements and the deployment of payloads from lunar orbit.

Background Information: The CEV SIM Bay could be used to deploy network stations on the lunar surface (see above) and make a wide variety of orbital geochemical, geophysical, mineralogical, photographic, and structural measurements that are critical to the outlined science objectives.

Landing Site and Other Operational Decisions

For a lunar outpost or any lunar mission, scientific input should be an integral component of the decision-making process for landing site targets, as well as exploration planning and execution.

Integration of Data Sets

Lunar data sets from all past missions, LRO, and future international missions should be geodetically controlled and accurately registered to create cartographic products that will enable landing site characterization, descent and landed operations, and resource identification and utilization through a variety of data fusion techniques.

Background Information: This recommendation grew out of the discussion of how to integrate the various data sets that will be returned from the Moon in the next 5–8 years as well as those previously collected.

Planetary Astronomy

Topics such as using the Moon or lunar orbital platforms to search for near-Earth objects and characterizing zodiacal dust should be integrated further into lunar exploration science planning.

Background Information: The area of planetary astronomy largely fell between the PSS and Astrophysics Subcommittee (APS) at this Workshop, because the PSS focused on lunar science and the Astrophysics Subcommittee focused primarily on astronomical targets outside the solar system.

Technology Developments

A lunar instrument and technology development program is needed to achieve several of the highest-ranked scientific objectives (e.g., exploration and sample documentation aids, long-lived 1-10 W power supplies; deployment of networks from orbit; sampling in permanently shadowed regions; development of robotically deployable heat-flow probes).

Background Information: Important technological developments are necessary in order to enable vital exploration science. Such technologies will not be lunar-specific, but will feed forward to Mars (and beyond). The specific technology development needs that were highlighted at the Tempe Workshop are listed below:

Technology Development Needs

- Imaging, ranging, position determination, and other aides to field exploration and sample documentation
- Long-lived (6-year life-time minimum) power supplies, especially in the 1-10 W range
- Interfacing of human and robotic field studies
- Hard vs. soft landing options (capabilities) for deploying instrument packages from orbit to set up networks
- Development of robotically deployable heat-flow probes
- Analytical capabilities in the field—efficient sample documentation and analysis by astronauts on EVAs and by robotic field assistants (e.g., hand-held laser Raman spectrometer, x-ray fluorescence spectrometer, etc.)
- Equipment development and systems integration for lunar fieldwork
- Automated instrumentation/equipment deployment capabilities
- Automated (robotic) sample return
- Technologies to sample, document samples, and make measurements in permanently shadowed environments

Sustained Scientific Input to Lunar Exploration Planning

Regular reviews of the major decisions that will influence the science outcome and legacy of lunar exploration should be carried out by the Council and its science subcommittees, and their findings and recommendations transmitted to the Council. Topics for such reviews should include:

- Options for full access to the Moon (low, mid, and high latitudes; near-side and far-side; polar)
- Pre- and post-landing robotic exploration opportunities and missions
- Options to mix human and robotic exploration
- Surface science experiments and operations at the human outpost
- Surface science experiments and operations during human sorties
- Mission planning
- Critical items in space hardware design, including:
    - delivery of science experiments to the lunar surface

- returned payload constraints
- upload of science (samples, data) from the lunar surface
- orbiting module science requirements (e.g., SIM bay)
- crew orbiting science operational requirements (e.g., portholes)
- mission control science requirements during operations

Authors

Clive Neal, Notre Dame University, Notre Dame, IN, *neal.1@nd.edu*
Charles Shearer, University of New Mexico, Albuquerque, NM, *cshearer@unm.edu*
Lars Borg, Lawrence Livermore National Laboratory, Livermore, CA, *borg5@llnl.gov*

APPENDIX 5: WORKSHOP FINDINGS PLANETARY SCIENCE SUBCOMMITTEE

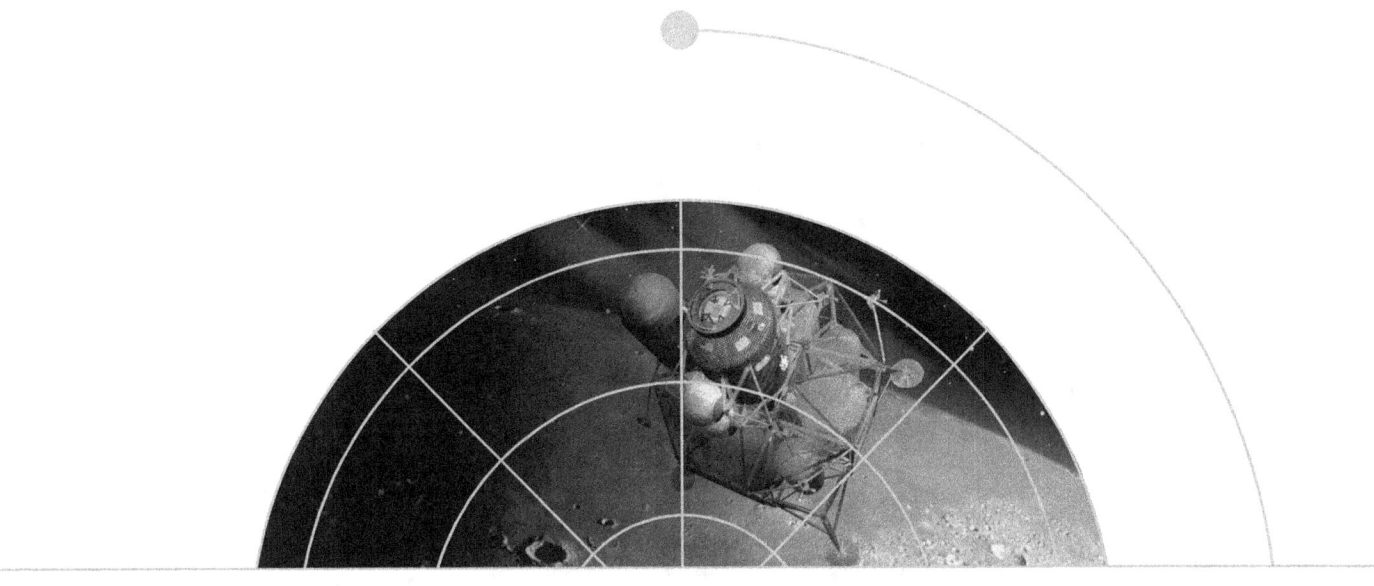

APPENDIX 6: DETAILED PROGRAM

NASA Advisory Council Workshop on Science

Associated with the Lunar Exploration Architecture

Detailed Program

February 27–March 2, 2007

Tuesday, February 27, 2007
Opening Plenary, Galleria Ballroom
Overflow Room: Encantada Ballrooom

8:00 a.m.	Opening Remarks	B. Jolliff, H. Schmitt
8:15 a.m.	The Vision for Space Exploration	M. Griffin
8:45 a.m.	SMD: Science Associated with the VSE	C. Hartman
9:00 a.m.	NRC Interim Report on the Scientific Context for Exploration of the Moon	C. Pieters G. Paulikas
9:30 a.m.	BREAK	
9:50 a.m.	ESMD General Welcome & Introduction	S. Horowitz
10:00 a.m.	Introduction of Global Exploration Strategy and Lunar Architecture Team	D. Cooke
10:10 a.m.	Global Exploration Strategy (including international and commercial components)	J. Volosin
10:50 a.m.	Overview and Status of the Lunar Exploration Architecture Team Activity	D. Cooke
11:00 a.m.	Science within the Lunar Architecture	L. Leshin
11:30 a.m.	LEAG TOP-SAT: Summary of Results and Science Objectives	J. Taylor
12:15 p.m.	LUNCH Subcommittee Science Discussion Overviews	
1:30 p.m.	Astrophysics Overview	D. Spergel
1:50 p.m.	Heliophysics Overview	J. Spann, H. Spence
2:10 p.m.	Planetary Protection Overview	J. Rummel
2:30 p.m.	Planetary Science Overview	C. Shearer
2:50 p.m.	Earth Science Overview	M. Ramsey
	The Lunar Earth Observatory Concept	P. Christensen
3:10 p.m.	BREAK	
3:30 p.m.	Subcommittee Breakout Sessions	

Tuesday, February 27, 2007
Astrophysics Subcommittee, Prescott Room

3:30 p.m.	Astrophysics Introduction: Review of STScI Meeting	Session Chair: D. Spergel
3:40 p.m.	Astrophysics Enabled by the Return to the Moon Report	M. Livio
4:30 p.m.	Astrophysics Theme 1: IR/Optical/UV Telescopes	
4:40 p.m.	Dirt, Gravity, and Lunar-Based Telescopes: The Value Proposition for Astronomy	D. Lester
5:30 p.m.	ADJOURN	

Tuesday, February 27, 2007
Earth Science Subcommittee, Palo Verde Room

	Breakout Session 1—NRC Decadal Survey Review	
3:30 p.m.	Earth Science Decadal Survey	M. Freilich
4:30 p.m.	ESD Road Mapping Process	B. Cramer
5:30 p.m.	ADJOURN	

Tuesday, February 27, 2007
Heliophysics Subcommittee, Payson Room

3:30 p.m.	Breakout Session 1—Heliophysics Science of the Moon	Session Chair: J. Spann
3:40 p.m.	Lunar Electromagnetic/Plasma Environment	B. Lin
4:10 p.m.	Determining Lunar Crustal Magnetic Fields and their Origin	J. Halekas
4:40 p.m.	The Lunar Wake as a Unique Plasma Physics Laboratory to be presented by J. Halekas	B. Farrell,
5:30 p.m.	ADJOURN	

Tuesday, February 27, 2007
Planetary Protection Subcommittee, Galleria Ballroom

3:30–5:30 p.m.	Breakout Session 1—Organic/Microbial Analyses + Experiments	Moderators: M. Voytek, C. Conley
	Theme 1: Overview of Life Detection Methods and Challenges	A. Steele
	Theme 2: Organic Measurements on the Lunar Surface: "Natural" and Planned Experiments	J. Dworkin
	Theme 3: The Urey Experiment with Planetary Protection Applications	J. Bada
	Theme 4: Organics in the Apollo Lunar Samples	C. Allen, J. Lindsay
5:30 p.m.	ADJOURN	

Tuesday, February 27, 2007
Planetary Science Subcommittee, Encantada Ballroom

	Breakout Session 1—Key Science Problems I	Moderators: J. Head, J. Taylor
3:30 – 5:30 p.m.	Theme 1: Impact History of the Inner Solar System	D. Kring, T. Swindle
	Theme 2: Exosphere	A. Stern (by telephone)
	Theme 3: Nature, Origin and Evolution of Volatile Polar Deposits	D. Lawrence, B. Bussey
	Theme 4: Indigenous Lunar Volatiles	M. Rutherford
5:30 p.m.	ADJOURN	

Wednesday, February 28, 2007
8:00 a.m.   Posters Display, Galleria Ballroom

Wednesday, February 28, 2007
APS, Prescott Room

8:30 a.m.	Breakout Session 2(a)—Astrophysics Theme 2: Talk, Radio	Session Chair: J. Mather
8:50 a.m.	The 21cm Background: A Low-Frequency Probe of the High-Redshift Universe	J. Hewitt
9:30 a.m.	Peering through the Dark Ages with a Low Frequency Telescope on the Moon	J. Burns
9:50 a.m.	Radio Wavelength Observatories and the Exploration Architecture	J. Lazio
10:15 a.m.	BREAK	
10:30 a.m.	Breakout Session 2(b)—Astrophysics Theme 1: High Energy Astrophysics	Session Chair: K. Flanagan
10:40 a.m.	High Energy Gamma-Ray and Cosmic-Ray Astrophysics on the Moon	R. Binns
12:00 noon	LUNCH	

Wednesday, February 28, 2007
ESS, Palo Verde Room

	Breakout Session 2(a)—Earth Science Decadal Survey Discussion	
8:30 a.m.	ESS Decadal Review Discussion: Which Activities could Map to a Future Lunar Earth Observatory/Earth Science Decadal Survey	All
10:00 a.m.	BREAK	
	Breakout Session 2(b)—A Lunar-Based Earth Observatory	Session Chair: M. Ramsey
10:15 a.m.	Introduction	M. Ramsey
10:30 a.m.	A Lunar Earth Observatory	P. Hamill
10:50 a.m.	Dual-use Earth Science and Lunar Exploration Missions	T. Freeman
11:10 a.m.	Science Observations from the Earth-Moon L1 Point	J. West
11:30 a.m.	Panel Discussion/Q&A (Ramsey, Freeman, Hamill, Christensen, West)	All
12:00 noon	LUNCH	

Wednesday, February 28, 2007
HPS, Payson Room

8:30 a.m.	Breakout Session 2(a)—Space Weather, Safeguarding the Journey	Session Chair: N. Schwadron

8:30 a.m.	Characterizing the Near Lunar Plasma Environment	T. Stubbs
9:00 a.m.	Dusty Plasma Issues on the Lunar Surface: Existing Observations and Required Future Measurements	M. Horanyi
9:30 a.m.	Space Weather Imaging from the Moon	D. Hassler
10:00 a.m.	BREAK	
10:15 a.m.	Breakout Session 2(b)—Space Weather, Safeguarding the Journey	Session Chair: N. Schwadron
10:30 a.m.	Space Weather Impacts on Robotic and Human Productivity	J. Mazur
11:00 a.m.	Characterizing Radiation Bombardment	J. Adams
11:30 a.m.	Systems on the Lunar Surface to Support of Space Weather	J. Davila
12:00 noon	LUNCH	

Wednesday, February 28, 2007
Joint PSS-PPS, Encantada Ballroom

8:30 a.m.	Breakout Session 2—Key Science Problems II	Moderators: C. Neal / C. Shearer
	Theme 1: Differentiation History of the Terrestrial Planets as Recorded on the Moon	L. Borg
	Theme 2: Structure and Evolution of the Lunar Interior	B. Banerdt, L. Hood
	Theme 3: Origin and Evolution of the Earth-Moon System	K. Righter (C. Shearer presenter)
10:15 a.m.	BREAK	
10:30 a.m.	Theme 4: Evolution of the Lunar Crust	B. Jolliff, L. Nyquist
	Theme 5: Science Associated with Resource Identification and Development	J. Taylor, M. Duke
	Theme 6: Surface Processes On Airless Planetary Bodies	L. Taylor
12:00 noon	LUNCH	

Wednesday, February 28, 2007
APS, Prescott Room

1:30 p.m.	Astrophysics Theme 3: Fundamental Physics and Astronomy	Session Chair: C. Hogan
1:40 p.m.	Fundamental Physics from Lunar Ranging	T. Murphy
2:20 p.m.	ALIVE: An Autonomous Lunar Investigation of the Variable Earth	M. Turnbull
2:50 p.m.	Science and Astrobiology from the Moon or near Moon.	N. Woolf
3:00 p.m.	BREAK	
3:30 p.m.	Astrophysics Theme 4: Astrophysics Quodlibet	Session Chair: M. Cherry
3:40 p.m.	Enabling Astrophysics at the Moon	Y. Pendleton

3:50 p.m.	A Large Optical/UV Serviceable Space Telescope	M. Postman
3:00 p.m.	Large Optics in Space	P. Stahl (2 min/ea)
	Poster previews	poster presenters

Wednesday, February 28, 2007
ESS, Palo Verde Room

	Breakout Session 3(a)—Land Imaging and Solid Earth Science	
1:30 p.m.	Introduction	B. Minster
1:40 p.m.	Visible/Near-Infrared Remote Sensing of Earth From the Moon	J. Johnson
2:00 p.m.	Land Surface Monitoring from the Moon	J. Mustard
2:20 p.m.	Thermal Infrared Data from the Moon: Hazards and Hot-Spots	M. Ramsey
2:40 p.m.	Lunar-based Large Baseline Synthetic Aperture Radar Interferometry of Earth	K. Sarabandi
3:00 p.m.	Panel Discussion/Q&A (Minster, Johnson, Mustard, Ramsey, Sarabandi)	All
3:15 p.m.	BREAK	
	Breakout Session 3(b)—Atmospheric Composition and Climate	
3:30 p.m.	Introduction	D. Jacob
3:40 p.m.	Lunar Observations of Changes in the Earth's Albedo (LOCEA)	A. Ruzmaikin
4:00 p.m.	Observations of Lightning on Earth from the Lunar Surface	S. Goodman
4:20 p.m.	Variability in Global Top-of-Atmosphere Shortwave Radiation	N. Loeb
4:40 p.m.	Panel Discussion/Q&A (Jacob, Herman, Goodman, Loeb, Ruzmaikin)	All
5:30 p.m.	ADJOURN	

Wednesday, February 28, 2007
HPS, Payson Room

1:30 p.m.	Breakout Session 3(a)—The Moon as a Historical Record	Session Chair: S. Suess
1:40 p.m.	Composition of the Solar Wind	S. Suess
2:00 p.m.	History of the Sun and Cosmic Radiation	K. Marti
2:20 p.m.	History of the Local Interstellar Medium—Cancelled	D. McKay
2:40 p.m.	History of Inner Solar System According to Lunar Cold Traps	D. Crider
3:00 p.m.	BREAK	
3:30 p.m.	Breakout Session 3(b)—The Moon as a Heliophysics Science Platform	Session Chair: A. Christensen
4:00 p.m.	Ionosphere/Magnetosphere Imaging	D. Gallagher
4:30 p.m.	The Moon as a Base for Solar Observations	G. Emslie
5:00 p.m.	Solar Observations Associated with the Return to the Moon	A. Title
5:30 p.m.	ADJOURN	

Wednesday, February 28, 2007
Joint PSS-PPS, Encantada Ballroom

	Breakout Session 3a—Implementation of Key Science into Lunar Exploration	Moderator: C. Shearer
1:30 p.m.	Theme 1: Important Scientific Sites on the Moon	J. Head
	Theme 2: Lunar Architecture's Plans to Provide Access to discussion Science Sites Cancelled	
	Theme 3: Geophysical Networks	C. Neal
	Theme 4: Importance of Sample Science and Sample Return	C. Shearer
	Sampling the SPA Basin: Some Considerations Based on the Apollo Experience	P. Spudis
	Theme 5: The Need for Integrating Planetary Protection Science and Technology	M. Race
3:15 p.m.	BREAK	
	Breakout Session 3a—"Implementation of key science into the exploration of the Moon and Mars"	N. Budden, L. Borg
3:30 p.m.	Theme 1: Human Surface Science	H. Schmitt
	Theme 2: Human-Robotic Combined Activities in Accomplishing Science	P. Spudis
	Theme 3: Linkages between the Moon and Mars	D. Beaty
	Theme 4: EVA Suit Competency for Science: Capabilities and Contamination	D. Eppler, J. Lindsay
	Theme 5: The AMASE Effort and Planetary Exploration	A. Steele
5:30 p.m.	ADJOURN	

Wednesday, February 28, 2007
Outreach, Ponderosa

3:40–5:30 p.m.	Outreach I	G. Kulcinski
4:00 p.m.	Lunar Exploration Outreach Program	K. Erickson
4:45 p.m.	Lunar Reconnaissance Orbiter Outreach	S. Stockman
5:30 p.m.	ADJOURN	

Wednesday, February 28, 2007
Poster session, Galleria Ballroom

6:00 p.m.	Poster session opens
	Cash bar, light snacks
8:00 p.m.	Poster session closes

Thursday, March 1, 2007
Plenary Session, Galleria Ballroom

8:15 a.m.	Introduction to Cross-cutting Topics, Thursday Agenda	B. Jolliff

Thursday, March 1, 2007
Exploration Science (Environment, Resources, Poles), Encantada

	Breakout Session 4—"Exploration Science"	Moderators: M. Duke, A. Steele
9:00 a.m.	Theme 1: ISRU Program Overview, including Timing	W. Larson
	Theme 2: Effects of ISRU on the Lunar Environment	R. Vondrak
	Theme 3: Space Weather	N. Schwadron
10:15 a.m.	BREAK	
	Theme 4: Physical / Chemical Properties and Potential Toxicity of Lunar Dust	L. Taylor
	Theme 5: Lunar Planetary Protection Testbeds and Life Support for Mars Exploration	J. Rummel
	Theme 6: Astrobiology and Lunar Exploration	A. Anbar
12:00 noon	LUNCH	

Thursday, March 1, 2007
Sun-Earth Interactions, Payson

	Breakout Session 4—Sun-Earth Interactions	Moderators: A. Christensen, K. Steffen
9:00 a.m.	Sun's Role in Climate Change	P. Goode
9:20 a.m.	A Possibility to Recover the Past Solar Constant (TSI) with the Moon	K. Steffen
9:40 a.m.	Imaging the Sun from the Moon	E. Deluca
10:15 a.m.	BREAK	
10:30 a.m.	Lunar JANUS Mission: An Exploration of the Earth and Sun	J. Herman
10:50 a.m.	Imaging Earth from the Moon	M. Turnbull
11:10 a.m.	Imaging Earth from the Moon	L. Paxton
	Panel Discussion (summarize key points)	All
12:00 noon	LUNCH	

Thursday, March 1, 2007
Lunar Dust Science, Palo Verde

	Breakout Session 4—Lunar Dust Science	Moderator: D. Winterhalter

9:00 a.m.	Introduction to Lunar Dust Science and Overview of the NESC Dust Workshop	D. Winterhalter
9:15 a.m.	Everything You Ever Wanted to Know about Lunar Dust	L. Taylor
9:35 a.m.	Interaction of Dust and Plasma on the Moon and Exosphere	T. Stubbs
9:55 a.m.	Measuring and Modeling the Plasma Environment	Z. Sternovsky
10:15 a.m.	BREAK	
10:30 a.m.	Dust Analysis at the Moon	Y. Pendleton
10:50 a.m.	Microwave Magnetic Properties of Dust and its implication for Geophysics and Cohesion	X. Yu
11:10 a.m.	Lunar Dust Distributions from Solar Infrared Absorption Measurements with a Fourier Transform Spectrometer	M. Abbas
11:30 a.m.	Autonomous Lunar Dust Observer for the Systematic Study of Natural and Anthropogenic Dust Phenomena on Airless Bodies	C. Grund
	Panel Discussion (summarize key points)	All
12:00 noon	LUNCH	

Thursday, March 1, 2007
Science Potentially Enabled, but not within Initial Scope, Prescott Room

	Breakout Session 4—Science Potentially enabled, but not within Initial Scope	Moderator: J. Mather
9:00 a.m.	Heliophysics Low Frequency Radio Astronomy	J. Kasper
9:20 a.m.	Synergies between Solar and Celestial Radio Astronomy"	J. Hewitt
9:50 a.m.	In-Space Capabilities Fostered by the Return to the Moon	H. Thronson
10:15 a.m.	BREAK	
10:30 a.m.	Costing Space and Lunar Missions	D. Ebbets
10:50 a.m.	Enabling Large Space Optics: SAFIR Human and Robotic Development	T. Espero
	Panel Discussion (summarize key points)	All
12:00 noon	LUNCH	

Thursday, March 1, 2007
Outreach, Ponderosa

9:00 a.m. - 12:00 noon	Outreach II	Session Chair: G. Kulcinski

Thursday, March 1, 2007
Plenary Session, Galleria Ballroom Cancelled (time given to final Subcommittee breakouts)

1:30 p.m.	Reports of Special Topics Breakouts	B. Jolliff
1:35 p.m.	Exploration Science	L. Taylor, A. Steele

1:50 p.m.	Sun-Earth Interactions	A. Christensen, K. Steffen
2:05 p.m.	Lunar Dust Science	D. Winterhalter
2:20 p.m.	Science Potentially Enabled, but not within Initial Scope	J. Mather
2:35 p.m.	Outreach	G. Kulcinski
2:50 p.m.	Introduction to Prioritizing the Science Objective Lists	B. Jolliff
3:00 p.m.	BREAK	
3:30 – 5:30 p.m.	Subcommittee Breakouts, Session 5	

Each of the Subcommittees will be asked to revisit the "Science Objectives Decomposition" matrix for their specific expertise, to use the remaining afternoon breakout sessions to prioritize the objectives, and to provide any additional comments or recommendations regarding the listed objectives and implementation issues. This is also the time for subcommittees to prepare a summary of findings and recommendations for the closing plenary session on Friday morning.

Thursday Afternoon Breakout Session Locations:
Astrophysics Subcommittee, Prescott Room
Earth Science Subcommittee, Palo Verde Room
Heliophysics Subcommittee, Payson Room
Planetary Protection Subcommittee, Galleria Ballroom
Planetary Science Subcommittee, Encantada Ballroom
Outreach Committee, Ponderosa Room

Friday, March 2, 2007
Closing Plenary, Galleria Ballroom

8:30 a.m.	Closing Plenary, Reports of Subcommittees, Lunar Architecture Team Remarks	B. Jolliff
8:40 a.m.	APS Findings, Recommendations	J. Mather
9:00 a.m.	ESS Findings, Recommendations	M. Ramsey
9:20 a.m.	HPS Findings, Recommendations	R. Torbert
9:50 a.m.	PPS Findings, Recommendations	J. Rummel
10:15 a.m.	BREAK	
10:30 a.m.	PSS Findings, Recommendations	S. Solomon
10:50 a.m.	Outreach Committee	G. Kulcinski
11:10 a.m.	Lunar Architecture Team Remarks	L. Leshin, LAT
11:30 a.m.	Conclude Workshop	B. Jolliff, H. Schmitt
12:00 noon	LUNCH	

1:30 p.m.	Synthesis Group—Payson Room	
	Synthesis committee reconvene	B. Jolliff, C. Neal
	Determine organization and format of final report	
	Synthesis committee writing assignments	
	Plan timeline for completion, review, and delivery of final product to NASA Advisory Council	
	Discuss Workshop Summary Report for EOS (AGU Newsletter)	N. Budden
	LEAG: Future activity related to the Lunar Exploration Architecture	C. Neal
4:00 p.m.	ADJOURN	

APPENDIX 6: DETAILED PROGRAM

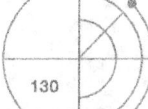

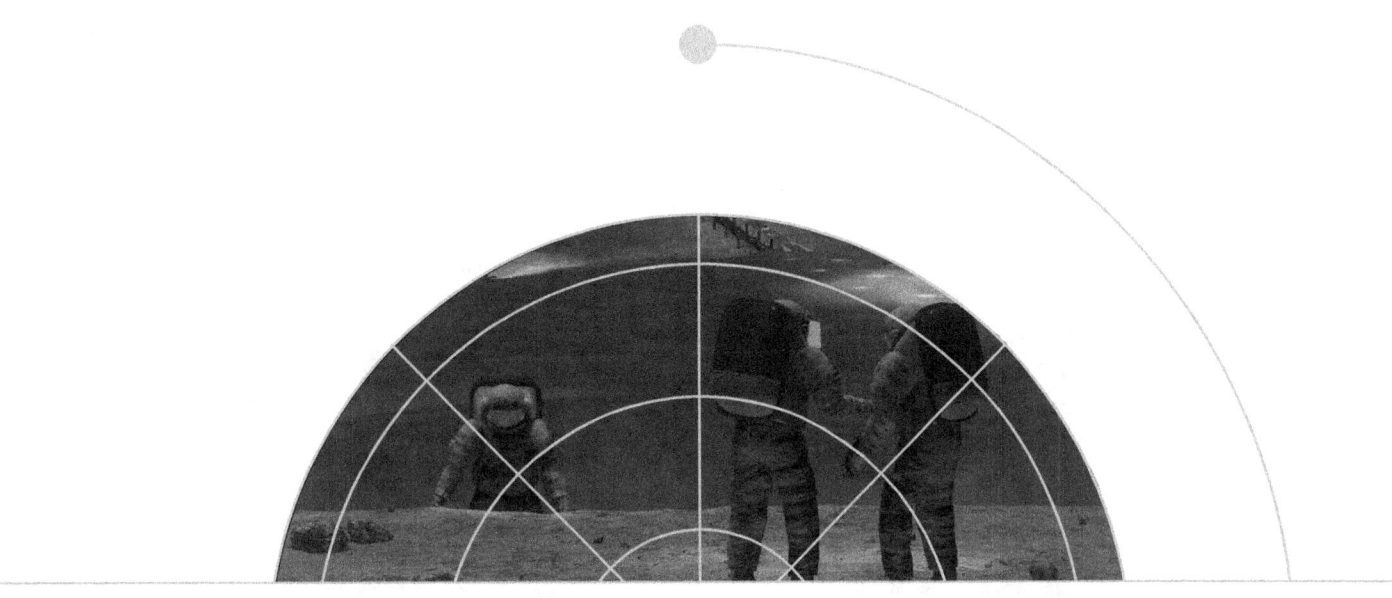

APPENDIX 7: LIST OF ACRONYMS

AGU – American Geophysical Union
AIRS – Atmospheric Infrared Sounder
APS – Astrophysics Subcommittee
APS – Active Pixel Sensor
ASTEP – Astrobiology Science and Technology for Exploring Planets
ASTER – Advanced Spaceborne Thermal Emission and Reflection Radiometer
ASTID – Astrobiology Science and Technology Instrument Development
AVHRR – Advanced Very High Resolution Radiometer
BRDF – Bidirectional Reflectance Distribution Function
CAPTEM – Curation and Analysis Planning Team for Extraterrestrial Materials
CCD – Charge-Coupled Device
CERES – Cloud and the Earth's Radiant Energy System
CEV – Crew Exploration Vehicle
COSPAR – Committee on Space Research
ENA – Energetic Neural Atom
ESAS – Exploration Systems Architecture Study
ESMD – Exploration Systems Mission Directorate
ESPA – Evolved Expendable Launch Vehicle Secondary Payload Adapter
ESS – Earth Science Subcommittee
ETM+ – LANDSAT Enhanced Thematic Mapper Plus
EVA – Extra-Vehicular Activity
FEAT – Field Exploration Activities Team
GEO – Geostationary Earth Orbit
GLAST – Gamma Ray Large Area Space Telescope
GOES – Geostationary Operational Environmental Satellites
GPS – Global Positioning System
GSFC – Goddard Space Flight Center
HiRISE – High Resolution Imaging Science Experiment
HPS – Heliophysics Subcommittee
INSAR – Interferometric Synthetic Aperture Radar
ISRU – In Situ Resource Utilization
ISS – International Space Station
JPL – Jet Propulsion Laboratory
JSC – Johnson Space Center
LAT – Lunar Architecture Team
LCROSS – Lunar CRater Observation and Sensing Satellite
LEAG – Lunar Exploration Analysis Group
LEO – Low-Earth Orbit
LIDAR – Light Detection and Ranging
LOCAD-PTS – Lab-on-a-Chip Application Development-Portable Test System
LRC – Langley Research Center

LRO – Lunar Reconnaissance Orbiter
LSAM – Lunar Surface Access Module
MEPAG – Mars Exploration Program Analysis Group
MIDP – Mars Instrument Development Program
MODIS – Moderate Resolution Imaging Spectroradiometer
MRO – Mars Reconnaissance Orbiter
MSFC – Marshall Space Flight Center
NOx – Nitrogen Oxide
NRC – National Research Council
NRL – Naval Research Laboratory
OMI – Ozone Monitoring Instrument
PIDDP – Planetary Instrument Definition and Development Program
PLSS – Portable Life Support Systems
PNT – Position, Navigation, and Timing
PPS – Planetary Protection Subcommittee
PSE – Passive Seismic Experiment
PSS – Planetary Science Subcommittee
RFI – Radio Frequency Interference
SAR – Synthetic Aperture Radar
SCFE – Science Capability Focus Element
SIM – Science Instrument Module
SMD – Science Mission Directorate
STScI – Space Telescope Science Institute
TES – Transition Edge Sensors
TSI – The Solar Constant
VSE – Vision for Space Exploration

www.ingramcontent.com/pod-product-compliance
Lightning Source LLC
Chambersburg PA
CBHW081238180526
45171CB00005B/462